The Practical Design of Advanced Marine Vehicles

Dr. Chris B. McKesson, PE

March 31, 2014

Contents

1 Summary & Purpose of this Textbook   **23**
1.1   Relationship of the Course to Program Outcomes   24
1.2   Prerequisites   25
1.3   A Note on Conventions   25
1.4   An Invitation to Edit   26

2 Navigating Without a Map   **27**

3 The Search For Speed   **31**
3.1   What is Fast? What is Speed?   32
3.2   Higher Froude Number means More Power   34
3.3   Hull Form vs Froude Number   36
    3.3.1   High Performance Monohulls   38
    3.3.2   Stabilized Monohulls   40
    3.3.3   Catamarans   42
    3.3.4   Wave Piercing Catamarans   43
    3.3.5   Hydrofoil Assisted Catamarans   44
    3.3.6   Hydrofoils   45
    3.3.7   Surface Effect Ships   47
    3.3.8   ACVs or Hovercraft   48
    3.3.9   Wing in Ground Effect or "WIGs"   51

4 The Sustention Space   **53**
4.1   The Sustention Triangle   53
    4.1.1   The Problem With The Sustention Triangle   54
4.2   The Sustention Cube   56
    4.2.1   First Axis: Static Lift or Dynamic Lift   56
    4.2.2   Second Axis: Aero- Lift or Hydro- Lift   57
    4.2.3   Third Axis: Powered or Passive   57
4.3   The Corners of the Sustention Cube   58

5 The Contents of the Sustention Cube   **59**
5.1   "Fast, Comfortable, and Cheap: Pick any two."   59
5.2   A Note On Scalability   60

5.3 The Advanced Marine Vehicles 62
   5.3.1 Passive Hydro Static (Buoyant) AMVs 63
   5.3.2 Passive Aero Static (Air Buoyant) AMVs 71
   5.3.3 Passive Hydro Dynamic (Dynamic Lift) AMVs . . . 71
   5.3.4 Passive Aero Dynamic (Dynamic Lift) AMVs 77
   5.3.5 Active Hydro Static (Powered Lift) AMVs 82
   5.3.6 Active Hydro Dynamic AMVs 92
   5.3.7 Active Aero-Static AMVs 93
   5.3.8 Active Aero-Dynamic AMVs 96
5.4 Summary of the AMVs in a single table 99

6 Hybrids and Weinblums   **101**
6.1 Hybrid Vehicles 101
   6.1.1 Missions And Speeds 102
   6.1.2 Speed And Lift 103
   6.1.3 Drag 106
   6.1.4 Drag Crises 108
   6.1.5 When Hybrids Work 111
   6.1.6 The V-K Gap: Physics Or Just Lack Of Imagination? 112
   6.1.7 Conclusion 113
6.2 What about Weinblums? Why must ships be symmetric? . 113

7 Performance Metrics   **117**
7.1 Von Karman / Gabrielli curve 118
7.2 The Value of Speed 119
   7.2.1 The Cost of Speed 120
   7.2.2 The Value of Speed 121
   7.2.3 Technology Affects Cost 122
   7.2.4 Cargo Affects Value 123
   7.2.5 Economics Affects Both 124
   7.2.6 What Does the Future Hold? 128
7.3 Kennell Transport Factor 129
   7.3.1 Transport Factor Defined: TF is L/D 129
   7.3.2 Transport Factor Decomposed 130
   7.3.3 Study of Size and Slenderness Effects 131
   7.3.4 Fuel Consumption - TF_{fuel} 132
   7.3.5 SFC effects 136
   7.3.6 Fuel Weight Fraction 136
   7.3.7 Emptyship Weight - TF_{SHIP} 136
   7.3.8 Conclusions on Kennell's Transport Factor 138
7.4 McKesson Parametrics 139
   7.4.1 The Sample Question 141

7.4.2   Major Parameters 143
7.4.3   Power Required - Ship Lift/Drag Ratios 144
7.4.4   McKesson "Observed Best Attainable" Curve 145
7.4.5   Weight of Power 146
7.4.6   Fuel Weight 147
7.4.7   Light Ship Weight 149
7.4.8   Putting it all together - A worked example 150
7.4.9   A range of Examples 151
7.4.10   The Design Space 152

8   Hydrostatic Balance   157

9   SWBS 051 - Resistance   159
9.1   The Resistance Components 159
9.2   Frictional Resistance 162
     9.2.1   Wetted Surface Variation 165
9.3   Wavemaking (Hull, not Cushion) 168
     9.3.1   Estimating Wavemaking Drag of a Single Slender Hull 169
9.4   Multihull Interference Drag 178
     9.4.1   Methods for predicting interference drag 181
     9.4.2   Model Testing Techniques 181
     9.4.3   Limitations 182
     9.4.4   Theoretical Interference Limits 183
9.5   Lift System Air Momentum Drag 184
9.6   Skirt Drag 185
9.7   Air Cushion Wavemaking 187
9.8   Spray and Spray Rail Drag 189
9.9   Appendage drag 189
9.10   Rules of Thumb 190

10 SWBS 070 - Hull Form Design   201
10.1 Catamaran hulls 201
     10.1.1 Catamaran hull form teleology 201
     10.1.2 Catamaran hull form parents 202
     10.1.3 Catamaran hull form development procedure 202
10.2 Trimaran Amas 203
     10.2.1 Trimaran Ama hull form teleology 203
     10.2.2 Trimaran Ama hull form parents 204
     10.2.3 Trimaran Ama hull form development procedure . . 205
10.3 SES Sidehulls 206
     10.3.1 SES Sidehull hull form teleology 206
     10.3.2 SES Sidehull hull form parents 209

10.3.3 SES Sidehull hull form development procedure . . . 209
10.4 SWATH Hulls 212
    10.4.1 SWATH hull form teleology 213
    10.4.2 SWATH hull form parents 214
    10.4.3 SWATH hull form development procedure 215

11 SWBS 070 - Ship Arrangement   **221**
11.1 General Arrangement 221
11.2 Aesthetics 227

12 SWBS 079 - Motions & Seakindliness   **243**
12.1 What is Unique About AMV Operations? 243
12.2 AMV-Unique Motions 244
    12.2.1 Corkscrewing 245
    12.2.2 Bow Diving 246
    12.2.3 Surface Suction & the Munk Moment 248
    12.2.4 Cobblestoning 249
    12.2.5 Plow-In 249
12.3 AMV Motions Analysis & Criteria 250
    12.3.1 AMV Motion Criteria 251
    12.3.2 Added Resistance 253
12.4 Motion Control for AMVs 253
    12.4.1 Modes of Control 253
    12.4.2 Effectors 254

13 SWBS 079 - Stability   **267**
13.1 Stability Curves for Multihulls 267
13.2 Hydrofoil Stability 268
13.3 SES Stability 269
    13.3.1 SES Static Stability 269
    13.3.2 SES Dynamic Stability 275
    13.3.3 SES Beam Sea Capsize 280
13.4 AMV Stability Criteria 281
    13.4.1 Intact Stability 282
    13.4.2 Damage Stability 284
13.5 AMV Intact Stability Tests 287

14 SWBS 100 - AMV Structures   **289**
14.1 Conventional Ship Load Cases 289
14.2 AMV Load Cases 289
    14.2.1 Longitudinal Bending Modes 290
    14.2.2 The Design Vertical Acceleration 292

14.2.3 Wave Height Limits 294
14.2.4 Design Pressures / Local Loads 297
14.2.5 Global Loads 299
14.3 AMV Load Cases Summary 303

15 SWBS 119 - Design of Air Cushion Skirts   **305**
15.1 Purpose and Types of Skirts 307
15.1.1 Virtual Skirts 309
15.1.2 Rigid Skirts 310
15.1.3 Inflatable Fabric Skirts 311
15.2 Basics of Inflatable Structures 314
15.3 Basic Design of SES Skirts 315
15.3.1 SES Bow Finger Skirts 316
15.3.2 SES Stern Bag Skirts 318
15.4 Skirt Forces 322
15.4.1 Internal forces 323
15.4.2 Attachment forces 323
15.4.3 Dynamic forces 324
15.5 Skirt Failures 324
15.6 Skirt Materials 325

16 SWBS 200 - Propulsors   **331**
16.1 The Propulsion Task - Required Thrust 331
16.1.1 Resistance Margin 331
16.2 Thrust Required 332
16.2.1 Hump Thrust Margin 332
16.2.2 Thrust Deduction 332
16.3 Propulsor types 333
16.3.1 Propellers 333
16.3.2 Waterjets 336

17 SWBS 200 - Propulsion Transmissions & Prime Movers   **357**
17.1 Transmitting Power to the Propulsor - AMV Unique Challenges 357
17.2 RPM Matching & Two-Speed Operations 359
17.2.1 Two Speed Gearboxes from ZF-Marine 359
17.2.2 Waterjets in Two-Speed Applications 360
17.3 Prime movers and their selection 361

18 SWBS 200 - Breguet's Range Equation   **363**

Contents

19 SWBS 500 - Lift Fan Systems   **367**

19.1 Cushion Air Demand - Estimating P & Q 367

    19.1.1 Air Flow Similitude 367

    19.1.2 The Hovergap Method for Air Demand 368

    19.1.3 Wave Pumping 369

19.2 Air Demand > Air Supply 372

19.3 Fans 101 374

19.4 Fan Scaling Laws 377

20 About the Author   **385**

List of Figures

2.1 Lewis and Clark, commemorated in US Postage 27

2.2 The author's summer residence, SUNDANCE at anchor in
     Canada 29

3.1 This small 20-knot speedboat is clearly "fast." 33

3.2 This 2-knot rowboat is clearly "slow." 34

3.3 Is this Washington State Ferry "Slow" or "Fast"? In numer-
     ical terms it is nearly the same speed as the speedboat, and
     yet in hydrodynamic terms it is as "slow" as the rowboat.
     This truth is captured through the naval architect's Froude
     Number. 35

3.4 Motoryacht DESTRIERO. Her Froude Number is approxi-
     mately the same as that of the speedboat in Figure 3.1 . . . 35

3.5 Speed and power data for a collection of vessels. 36

3.6 The same vessels as the preceding Figure, but now presenting
     Specific Power versus Speed 37

3.7 The same vessels as the previous two Figures, but now pre-
     senting Specific Power versus Non-Dimensional Speed (Froude
     Number). 37

3.8 The same data as Figure 3.7, colored to show hull type . . . 38

3.9 Most surface combatants are classed as High Performance
     Monohulls 39

3.10 The MDV3000 Fast Ferry JUPITER, built by Fincantieri . 40

3.11 CABLE & WIRELESS ADVENTURER, built for the 1998
     around-the-world record. (Photo from www.solarnavigator.net)
     41

3.12 Photograph of the trimaran ferry BENCHIJIGUA EXPRESS,
     built by Austal Shipyards. 42

3.13 The 122m Stena HSS 1500 catamaran ferry, in service on the
     Irish Sea. 43

3.14 The Washington State Ferry catamaran SNOHOMISH (now
     retired.) 44

3.15 Australian naval architect Phil Hercus' first commercial wave-
     piercing catamaran. 45

3.16 The Theater Support Vessel SPEARHEAD, a military Wave Piercing catamaran, after the pattern invented by Phil Hercus. 46

3.17 The Argentine ferry PATRICIA OLIVIA II 46

3.18 A hydrofoil-assisted catamaran. Photo from www.foils.org . 47

3.19 US Navy "PHM" hydrofoil patrol craft 48

3.20 A commercial Boeing JetFoil 49

3.21 Norwegian Cirrus 120P class Surface Effect Ship ferry, circa 1995. 49

3.22 Norwegian Navy SKJOLD SES patrol craft, circa 2000 . . . 50

3.23 The US Navy SES 100B 50

3.24 The English SR.N-4 commercial hovercraft, which served across the English Channel for over 30 years. 51

3.25 The Caspian Sea Monster - a Wing-in-Ground Effect (WIG). 52

4.1 The Sustention Triangle, including illustrations of some of the ship types at various points therein 54

4.2 The Sustention Cube, the author's alternative model of the AMV design space. This model offers broader applicability by covering more of the design space than the Sustention Triangle. 56

5.1 The first of the INCAT 74m Wave Piercing Catamarans - HOVERSPEED GREAT BRITAIN, who then held the record for the TransAtlantic Crossing. 65

5.2 The Earth-Race trimaran, the most exotic looking trimaran I have come across 66

5.3 The Austal trimaran ferry BENCHIJIGUA EXPRESS. Photos from www.austal.com 67

5.4 Austal's US Navy Littoral Combat Ship "LCS 2" in drydock 68

5.5 The parts and nomenclature of a SWATH. Picture taken from www.swath.com 69

5.6 US Navy T-AGOS 19 70

5.7 US Navy T-AGOS 19 ̄ 71

5.8 US Navy T-AGOS 19 72

5.9 SWATH Pilot Vessel from German shipyard Abeking and Rasmussen 73

5.10 Four-hulled SWATH variant SLICE 74

5.11 A Surface-Piercing hydrofoil produced by Rodriquez 76

5.12 A hydrofoil craft having fully-submerged foils. (The foils are visible below the sea surface in this photo) 76

5.13 The Prototypical Wing In Ground Effect 78

5.14 The Caspian Sea Monster 78

5.15 Rozhdestvensky's illustration of the benefit of ground effect
      upon vehicle Lift/Drag ratio 79
5.16 This illustration of the forces on a tunnel boat (from www.screamandfly.com)
      highlights the fact that these craft too are WIGs 80
5.17 The Reverse-Delta configuration preferred by Anton Lippisch 81
5.18 A tandem-wing WIG craft from Gunther Jörg 81
5.19 A simple schematic section illustrating the defining parts of
      a hovercraft. 85
5.20 Sir Christopher Cockerel 86
5.21 One of the first hovercraft, the Saunders-Roe N-1 (SR.N-1)
      Note the absence of fabric skirts as are used today. 87
5.22 The SR.N-1 in overwater operation. Note the large amount
      of spray created. 88
5.23 The Saunders-Roe N-4 (SR.N-4) commercial ferry. Note the
      greatly reduced spray compared to the SR.N-1, due largely
      to the use of fabric skirts of a design which is still current. . 89
5.24 A Russian AIST class amphibious military hovercraft, gen-
      erally equivalent to the USN LCAC 90
5.25 A Russian LEBED Class ACV 90
5.26 The largest hovercraft in the world, the Russian POMORNIK
      Class at 555 tonnes 91
5.27 A commercial hovercraft, exploiting the hovercraft's amphibi-
      ous capability in order to operate in ice. 92
5.28 The USN LCAC hovercraft 93
5.29 This picture of an LCAC clearly shows the role a hovercraft
      can have in shallow-water operation 94
5.30 This picture shows the ultimate in shallow-water: An LCAC
      on the beach, with the air cushion turned off. Note the
      deflated skirt visible around the perimeter of the craft. . . 95
5.31 The two 100-ton testcraft SES 100A and SES 100B 95
5.32 The SES 100A, the waterjet driven testcraft 96
5.33 The SES 100B, the propeller-driven testcraft 96
5.34 A commercial SES ferry from Norway 97
5.35 The Norwegian Navy SES Patrol Boat SKJOLD 97
5.36 The Boeing hydrocopter at rest 98
5.37 The Boeing hydrocopter under way 98

6.1   Power versus dynamic lift fraction for the example given in
      text 108
6.2   Vehicle weight versus dynamic lift fraction for the example
      given in the text 109
6.3   A bad planing boat but a good hydrofoil? 111

6.4   A sketch of a grapevine, or "weinblum." Note how the leaves
      are staggered port-starboard-port-starboard etc. 114
6.5   Herr Dr. Georg Weinblum 115
6.6   A plot of the wave pattern from a Weinblum hull, consisting
      of two identical hulls staggered longitudinally 116

7.1   Theodore von Karman 119
7.2   Von Karman's 1950 graph of Transport Efficiency [1] 120
7.3   Von Karman data collected by a class of undergraduates . . 121
7.4   The unarguable truth of the Cost of Speed 122
7.5   The conceptual sketch of the Value of Speed 123
7.6   The cost of speed depends upon the technology selected.
      (NB: Lines depicted are notional only.) 124
7.7   About fifty years of "value of time" data for people, corrected
      for inflation, from the Wikimedia Commons. 125
7.8   Nearly a century of "Cost of Energy" data, to compare with
      the previous graph 126
7.9   The value of time for goods (interest rates) for 50 years of
      US history (Source: DollarDaze.org) 127
7.10 Kennell's TF trendline, from [2] 131
7.11 TF Trendline proposed by Dr. Julio Vergara (Chile)(Private
      Communication) 132
7.12 Kennell's experience data for small fast ships. Note that not
      all of them are able to arrive at "State of the Art" performance 133
7.13 Kennell's data on the effect of slenderness, from [3] 133
7.14 Kennell's graph of the effect of size upon attained TF . . . 134
7.15 Kennell's historical data on TF fuel trends 135
7.16 Kennell's plot of the relationship between propulsion tech-
      nology and TF_{fuel} 137
7.17 Kennell's finding on the proportion of TF devoted to fuel,
      as a function of speed and range 138
7.18 Kennell's finding of the relationship between ship weight,
      cargo weight, and SHP. Note here that W_{cargo}/SHP is in
      units of Long Tons per Horsepower. Note also that the W_{ship}
      on the x-axis refers to the weight of the empty ship, not the
      weight of the total ship. 139
7.19 Kennell's graphic depiction of the nature of Deadweight Den-
      sity for different ship types 140
7.20 Low density payloads tend to demand higher values of light-
      ship weight fraction 141
7.21 High speed ships follow the same trend 142
7.22 Even aircraft follow the same trend! 143

7.23 Kennell's curve showing the effect of size upon TF 145
7.24 McKesson's Observed Frontier of ship TF. 147
7.25 Propulsion Gas Turbine Engines, SFC versus Power, Current and Future Engines 148
7.26 Propulsion Gas Turbine Engines, SFC versus Year of Introduction, Current and Future Engines 149
7.27 A worked example of a Very Simple Model for the HSSL mission 153
7.28 How much must the displacement grow, to obtain the targeted value of cargo? 154
7.29 Map of Ship Size versus k_{1356}. This shows the impact of specifying heavy solutions for structure and auxiliary systems, or contrariwise the incentive for developing lightweight structure and auxiliary systems. (Corresponds to 5000 LT cargo, 43 kts, 5000 nmi range, TF per Observed Best Attainable Curve, Weight of Power = 10 lbs / shp.) 155

9.1 Drag components of a 70m catamaran, from Faltinsen [4] . 162
9.2 Drag components of a 40m SES, from Faltinsen [4] 163
9.3 C_f Curve Comparison, from Faltinsen [4] 164
9.4 AMV design often feels like navigating using maps like this 165
9.5 A reproduction of Faltinsen's reference on Running Sinkage of a catamaran [4] 167
9.6 Kolazaev's figure for $Kf(Fn)$ 167
9.7 The wetting tapes (the two gold strips) fitted to the HSSL model to measure wetted girth. Three such sets of tapes were installed at different stations along the length of the model. 168
9.8 The dynamic wetted surface variation with speed as measured on the HSSL model 169
9.9 Wave pattern and distribution of wave pattern resistance as estimated by Michell's integral, from Lazauskas and Tuck [5] 173
9.10 Lundgren SSPA series parameters compared to other series 174
9.11 Contours of Residuary Resistance Coefficient for $B/T =3$ $CB = 0.40$ from the Lundgren series [6] 191
9.12 Total Resistance Coefficient for six Arrow Trimaran configurations, from Lazauskas and Tuck [5] 192
9.13 CFD and model test results, for a 2009 study of the effect of longitudinal position of side hulls on trimaran residuary resistance 193
9.14 Comparison of the free surface behind trimaran 5651 in Experiment 5 (left) and Experiment 9 (right) at Froude Number = 0.34 193

9.15 Total resistance of optimized one-tonne generalized trimarans [5] 194

9.16 Doctors' geometry definition sketches for a stern seal (left) and a bow seal (right) 194

9.17 An SES stern seal exactly corresponding to Doctors' definition sketch 195

9.18 The wave pattern caused by a rectangular constant-pressure patch 195

9.19 Newman and Poole cushion wave drag parameter 196

9.20 Doctors' figure showing the Newman and Poole instability, and the smoothing accomplished by introducing parameters alpha and beta 197

9.21 Doctors' pressure smoothing parameters 198

9.22 Doctors' results for cushion wavemaking drag 199

9.23 A US Navy result for total drag of an 8,000 ton SES as a Function of Speed and L/B ratio [7] 200

10.1 Saunders' guidance for the selection of desired Cp and Fatness Ratio 203

10.2 Gives some depiction of the form of Ama preferred by Dr. Tony Armstrong 207

10.3 A depiction of the SWATH-like Amas preferred by Dr. Igor Mizine 208

10.4 Typical variation in SWATH ship heave response at low speeds as a function of tuning factor. (SNAME) 213

10.5 NATO Standard sea state definitions 214

10.6 FEffect of ship speed on wave encounter period in head seas 215

10.7 High Cp / Low Speed parent SWATH T-AGOS 215

10.8 High Cp / Low Speed Parent: SWATH T-AGOS-B 216

10.9 Low Cp / High Speed Parent: SWATH 5972 216

10.10 Lamb's definition sketch for angle Beta 218

10.11 Lamb's definition sketch for angle Alpha 219

11.1 Galapagos Islands tourboat ANAHI, showing the standard arrangement of a catamaran 223

11.2 KAIMALINO, pioneering an unusual arrangement approach 224

11.3 The Canadian PacifiCat fast ferry. The bridge is not the top deck, but the one right below it. 225

11.4 A detail of a Pacificat, showing the overhanging bridge wing 225

11.5 A luxury hotel atrium. Given the smooth ride of a SWATH ship, why not use a configuration like this? 226

11.6 A four-story atrium, with proportions that might fit many
AMVs 229
11.7 A hotel atrium. Could this be used on a small catamaran? . 230
11.8 RADISSON DIAMOND, a SWATH cruise ship 230
11.9 A Low-Res section through RADISSON DIAMOND 231
11.10RADISSON DIAMOND Stern View 232
11.11The STENA HSS 1500 fast ferry 232
11.12USN SWATH T-AGOS 233
11.13Monohull T-AGOS 234
11.14SWATH T-AGOS 235
11.15FREDERICK CREED, a small SWATH 236
11.16Arrangement drawings of the INCAT K-50 car ferry 237
11.17Austal's illustration to compare the flight deck size on an
AMV versus several monohulls 238
11.18SEA SHADOW 238
11.19SEA SHADOW from above. Note the lower hulls that are
dimly visible under the water, forward. 239
11.20VICTORIA CLIPPER IV 240
11.21A counter example, with too many lines going in too many
different directions 240
11.22STARSHIP EXPRESS 241

12.1 The limiting wave height table for the X-Craft, at 1400
tonnes and below, in head seas 245
12.2 The X-Craft 246
12.3 The relationship (in deep water) between wave speed (Celerity $= \sqrt{(gL/2\pi)}$) and wave length 247
12.4 MCA Photo sequence of model tests of a catamaran bow dive 248
12.5 The Plow-In process, from [8] 250
12.6 O'Hanlon & McCauley criteria for motion sickness, as presented in ISO 2631 252
12.7 Ugo Conti's Spider Boat. Photo from SFGate website. . . . 255
12.8 A Maritime Dynamics T-foil 257
12.9 An MDI Trim Tab, 3-D view 258
12.10A trim tab, profile view, showing the pressure effect on the
bottom. 258
12.11An MDI Interceptor, 3-D view 259
12.12An Interceptor profile view, showing the pressure effect on
the bottom. 260
12.13The steering forces due to a KaMeWa-style steering and reversing suite 262

12.14The steering forces due to a Rams-Horn style steering and
reversing suite 263

12.15An LCAC Class ACV 264

12.16A blow-up of the LCAC's propulsion nozzle, with the rudders
marginally visible behind them. 265

12.17A blow up of the LCAC's bow thrusters (the snorkel-like
structures near the center of the photo.) 266

12.18The many appendages of the SES 100A 266

13.1 Monohull Stability - G below B 269

13.2 Monohull Stability - G above B 270

13.3 Trimaran Stability - G above B 271

13.4 Catamaran Stability - G above B 272

13.5 Taken from a forgotten site on the internet, this graphic does
an excellent job of contrasting the stability of three types of
craft. 273

13.6 Another internet-harvested graphic, depicting the situation.
The condition of a trimaran is like that of a monohull with
G above B. 274

13.7 Blyth's illustration of the balance of righting forces for an
SES on cushion. 275

13.8 Blyth's illustration of the effect of emergence of the sidehull
as an SES heels 276

13.9 Forces acting on an SES in a high speed turn 277

13.10The roll moments associated with the forces in Figure 13.9 277

13.11The effect that roll angle has upon the moment induced by
the planing force resultant 278

13.12The effect of VCG on Roll Moments 279

13.13Effect of Hull Form on Critical KG 280

13.14Typical SES capsize sequence in Beam Seas 281

13.15Lewthwaite's 1986 guidance on form parameters to avoid
capsize. The black spots were tested craft. The large grey
spots were designs that were then under evaluation. The
validity of this curve has not been proven. 283

13.16A USCG illustration based on the Assumption that Max RA
occurs above 35 degrees of heel. 284

13.17Illustrating the assumption that most righting arm curves
are positive to at least 90 degrees 285

14.1 DNV "Crest Landing"condition, equivalent to hogging . . . 291

14.2 DNV "Trough Landing"condition, equivalent to sagging . . 291

14.3 The selection of Acceleration Factor as a function of Service
Restriction Notation and Ship Type 293
14.4 Longitudinal distribution factor for design vertical acceleration 294
14.5 The spreadsheet used to calculate Figure 14.6 295
14.6 A speed / wave height relationship selected to yield constant
design acceleration 296
14.7 A practical limiting wave height curve overlaid on Figure 14.6 296
14.8 The decrease of slamming pressure toward the stern 298
14.9 Figure 199 - Longitudinal variation of wet deck slam pressure 299
14.10 DNV's formula for Sea Pressure 300
14.11 Sea Pressure longitudinal distribution factor k_s, a function
of block coefficient 301
14.12 Transverse bending moments and shear force 302
14.13 The pitch connecting moment, decomposed into Mp and Mt 303

15.1 A hovercraft skirt system, simply to illustrate the complexity
of this engineered product 306
15.2 Yun & Bliault [8] description of the governing equations for
peripheral jets. 310
15.3 One type of inflated skirt. 313
15.4 A Pericell and Bag (or Jupe and Bag) skirt system 314
15.5 The finger skirt (right) explained as a derivative case of a
single curtain skirt. 315
15.6 A bag-and-finger skirt system 316
15.7 Basics of inflatable structures 318
15.8 Drawings of generic SES bow-finger geometry 319
15.9 A two-lobed SES bag-type stern seal 319
15.10 Definition sketch for a simplified case of the geometric balance of a stern bag seal 320
15.11 A bolt-rope type skirt attachment scheme 325
15.12 A bolt-rope scheme involving a bolted clamping system . . 326
15.13 Mechanical hinge attachment concepts 326
15.14 Bolted attachment of skirt segments, with bolts protected by
anti-chafe rings (see next figure.) 327
15.15 An anti-chafe collar 327
15.16 Attaching the fingers to the bag 328
15.17 An SES bow skirt, where the wear at the tips of the fingers
due to flagellation is clearly visible 329
15.18 Showing the afloat detachment of two bag segments from a
three-lobed stern seal 329
15.19 Data table from Yun & Bliault describing two skirt fabrics
available in China [8] 330

15.20 Data table from Yun & Bliault describing skirt materials and life from some built SES and ACV [8] 330

16.1 335
16.2 Newton-Rader series blade section shapes 336
16.3 Performance characteristics of the Newton Rader series propellers. From [4] 337
16.4 Twin surface-piercing propellers on a race boat 338
16.5 A Surface-Piercing Propeller test rig, which illustrates the major parameters of the SPP 338
16.6 A photo of the air cavity behind a surface piercing propeller 339
16.7 Rose & Kruppa data for a surface piercing propeller with P/D=1.75, 12° shaft angle 339
16.8 Theoretical waterjet jet efficiency, for practical values of JVR and wake fraction, from [9] 341
16.9 An early waterjet based on a centrifugal-type pump 342
16.10 An early waterjet based on an axial-type pump 342
16.11 A textbook illustration of a centrifugal pump 343
16.12 Textbook illustration of an axial pump 344
16.13 A Cordier diagram of pump regimes 345
16.14 A mixed-flow waterjet 345
16.15 A mixed-flow waterjet 346
16.16 KaMeWa S-Series units, relating size (model number) to power 346
16.17 Geometry of the KaMeWa S-series 347
16.18 348
16.19 A profile of a waterjet inlet illustrating the pressures experienced on the boundary, from [4] 348
16.20 Surface pressures in a flowing waterjet inlet 349
16.21 A KaMeWa quotation for a specific project, involving quadruple size 153 waterjets 350
16.22 Illustrates the case of a craft entering the cavitation zone for a brief period for an event such as hump transit. From [4] . 351
16.23 Relationship between power, RPM, and speed for a waterjet 353
16.24 Attained waterjet performance values for one design project 354
16.25 A Wartsila jet, clearly showing the location of the thrust bearing 355

17.1 A typical AMV diesel engine power map 360
17.2 Two-speed gearboxes available from ZF Marine 361
17.3 Gear ratios available on the ZF two-speed gears 362

19.1 Stylized illustration of the hovergap for an ACV (top) and
       an SES (bottom) 369
19.2 Air flow demand data for a collection of hovercraft 370
19.3 The data from Figure 19.2, plotted showing an apparent sen-
       sitivity of Flow to Speed 370
19.4 A crude sketch of an SES profile, showing the volume of the
       cushion that must be refilled with air between the passage
       of a crest and a trough. 371
19.5 The desired lift fan Pressure / Flow characteristic 372
19.6 The shape of a real fan's pressure / flow characteristic . . . 373
19.7 A real SES lift fan. The curve for "FSP" is the fan static
       pressure in inches water gage, plotted versus the flow in cfm
       x 10,000. Other curves give efficiency and power consumed
       by this fan. 374
19.8 Syracuse University slide on the types of Fluid Movers . . . 376
19.9 Depiction of the difference between axial and centrifugal
       aeromachinery 377
19.10 A mechanical engineer's illustration of two axial flow machines 378
19.11 This turbocharger shaft shows two mixed-flow machines, one
       (the turbine) to extract energy from the exhaust gas and the
       other (the compressor) to impart energy into the inlet flow 379
19.12 Howden Buffalo fan product ranges 381
19.13 A given fan design, in two different sizes to yield two different
       P/Q curves 382
19.14 The same two fans as in Figure 265, but when plotted non-
       dimensionally revealed to be the same turbomachine 383

List of Tables

5.1 The author's subjective assessment of various AMV hull forms against five performance parameters. A high number indicates better performance. 60

14.1 A simple parametric look at the values given by DNV's formula for Design Vertical Acceleration 292

18.1 The effect of the Breguet range calculation 365

19.1 Three different parent fans all scaled to $P = 6.03$ kPa & $Q = 200$ cms 380

21

1 Summary & Purpose of this Textbook

This text is a written version of University of New Orleans course "NAME 4177" in the School of Naval Architecture and Marine Engineering, College of Engineering. The course is a 13-week twice-a-week elective, at the Senior undergraduate level. This text will provide an introductory familiarity with the naval architecture of Advanced Marine Vehicles, with particular emphasis on Catamaran, SES and SWATH types. It is assumed that the students have a working familiarity with the naval architecture of conventional ships, and thus this course emphasizes the differences between conventional-ship design and AMV-design. The course is focused on early-stage design, providing the tools for preliminary ship sizing in order to evaluate whether the AMV is the appropriate ship type for the mission. The course will include discussion of the particular features and benefits of the major AMV types, so that you can decide when one AMV type might be preferable over another. The course will begin with an overview of the types of AMVs. This is followed by discussions of each of the "nodes" of the ship design spiral, e.g. Resistance, Propulsion, Structural Design, Arrangement, Maneuvering, etc. At the conclusion of this course the student should be able to:

- Recognize the different types of Advanced Marine Vehicles

- Know the specific features (Pros and Cons) of the differing AMV types

- Select an AMV type for a given mission

- Perform initial sizing of the selected AMV

- Estimate the resistance of the selected AMV

- Size the Lift System of an SES or ACV

- Perform a weight estimate for a multihull (including Catamaran, Trimaran, SWATH, and SES)

- Understand the structural load mechanisms peculiar to AMVs

- Pursue weight-reduction technologies that may be essential to AMVs

- Evaluate a newly-proposed AMV type for merit and feasibility

- State the nature and magnitude of the AMV's environmental impact

- Know where to look for specialist technical resources, including literature and people

- Know where your weaknesses lie for follow-on design phases, so that you can solicit the needed specialist help

Appropriate to being an overview type of course at the undergraduate level, this course does not provide a detailed treatment of any of the hydrodynamic or mechanical dynamic nuances of high speed vessel design. Instead the course presents design lanes and overall guidance, such that a practitioner can execute a reasonable early-stage design. Tackling of specific detailed problems that may come up within such a design exercise may require recourse to more detailed texts, and appropriate references and citations are provided herein. Finally, let me state right up front that this work is not "definitive." Many fine thinkers have written important works on this subject, and a truly "definitive" book would probably have to include those predecessors verbatim. Rather, this work is intended to be a usable, foundational work, suitable for a single-term course of study, and as a reference that will direct the advanced student to those more detailed works upon which I have drawn.

1.1 Relationship of the Course to Program Outcomes

UNO NAME 4177 contributes to the following standardized outcomes, as defined by ABET, Inc., the recognized accreditor for college and university programs in applied science, computing, engineering, and technology. ABET is a federation of 29 professional and technical societies representing these fields. For more information on ABET and the accreditation services they provide visit www.abet.org. ABET Outcomes:

1. a) _X_ An ability to apply knowledge of mathematics, science, and engineering.

    b) _X_ An ability to design and conduct experiments, analyze and interpret data.

    c) _X_ Ability to design a system, component, or process to meet desired needs.

d) ___ Ability to function on multi-disciplinary teams

e) _X_ Ability to identify, formulate, and solve engineering problems

f) _X_ Understanding of professional and ethical responsibility

g) _X_ Ability to communicate effectively

h) _X_ Understand the impact of engineering solutions in a global and societal context

i) _X_ Recognition of the need for, and ability to engage in life-long learning

j) _X_ Knowledge of contemporary issues

k) _X_ Ability to use the techniques, skills, and modern engineering tools necessary for engineering practice

l) _X_ Ability to apply probability and statistical methods to naval architecture and marine engineering problems

m) _X_ Basic knowledge of fluid mechanics, dynamics, structural mechanics, material properties, hydrostatics, and energy/propulsion systems in the context of marine vehicles

n) ___ Familiarity with instrumentation appropriate to naval architecture and marine engineering

1.2 Prerequisites

Senior standing in the School of Naval Architecture and Marine Engineering. It is assumed that the student has a journeyman understanding of conventional naval architecture in all of its disciplines: Hull forms, stability, resistance and powering, ship strength, ship motions, ship maneuvering and control, etc.

1.3 A Note on Conventions

Note that this course outline uses the USN SWBS numbering convention. A SWBS manual ([10]) is provided in the course reference materials. The UNO version of the course includes a lecture on SWBS, but this lecture has not been included in this text version of the course.

In the same vein, I have endeavoured to adhere to the SI system of units in this text and their abbreviations. In particular, I invite the reader to note that the accepted abbreviation for the metric ton is "t" and not "MT" or any other symbol.

1.4 An Invitation to Edit

This book was originally hosted on the author's website as a wiki, and students were invited to directly edit the text and contents of the book. For a variety of reasons I have moved into this printed form, but this does not change the fact that the work might be continuously improved by its readers and users.

To this end, I invite any interested party to suggest corrections, additions, or deletions from this work.

The author may be reached at chris@mckesson.US

2 Navigating Without a Map

Figure 2.1: Lewis and Clark, commemorated in US Postage

Lewis and Clark are household names in my home state of Washington, where we pride ourselves on being "discovered" by these intrepid explorers. These two brave men pushed into what was then unknown territory. My wife and I have driven over the mountains of the west, and we often comment on what it must have been like to climb those mountains on foot, never knowing what would be seen on the other side. What is it like to navigate without a map?

AMV designers are often in the situation of navigating without a map. Our monohull brethren are able to look at myriad examples of prior art. Indeed, in undergraduate courses in conventional ship design we begin the design process by collecting a database of similar ships, and generating our ship's characteristics by gentle interpolations and extrapolations within that "mapped" design space. In the case of AMV design we are often left without such a map, and we must have recourse to more fundamental tools.

Let me shift my metaphor from Lewis and Clark to one of "learning to drive in America." If my course in AMV design may be likened to a course in driving, then this first lecture in the course would be like having the driving instructor begin the session by handing out a compass, a hatchet or machete, and other tools of the backwoodsman. We are going to go where there are no roads. Learning to "drive" the process of AMV design means we have to also learn how to build our own road through virgin territory.

We are Naval Architects in the spirit of Lewis and Clark. The design of AMVs is, by definition, unknown territory, and practitioners in this field are explorers. Good exploration takes different skills than using a trail already blazed by somebody else.

Figure 2.2 is a photograph of my primary residence - a 1968 Columbia 36 sailboat SUNDANCE. My wife and I have lived on our boat on-and-off for the better part of ten years. Living on a boat has many challenges, and not all of them are the ones that are well-documented and described in the literature of the field. Take, for example, the practical question "How do you mount a Christmas tree on a sailboat?" West Marine doesn't sell Christmas tree stands. And the typical Wal-Mart tree stand will not be a good solution in the dynamic environment of a boat. So how do we mount our Christmas tree? We have no "prior art" to draw upon. We have no guidance on what the "tried and proven" solution is. We are forced into unknown territory, equipped only with our basic tools and our wits, and we are free to invent our own solution.

In similar fashion, in the case of AMV design it is rare that we have a systematic series of prior art to draw from. So just like the problem of "How do you mount a Christmas tree on a sailboat?" we are free to invent new solutions, instead of doing it "the same way the last fella did it." Of course, with the arrogance of the AMV designer, we like to respond that this gives us our new mantra: "Don't do it the way the last fella did, do it right instead." At this point I need to hasten to repeat a counterbalancing maxim: Despite our freedom to do things a new way, it is simultaneously important to avoid gratuitous innovation - innovate only when needed. Mr. Bob Colwell of INTEL said it well: "Creativity is a poor substitute for knowing what you are doing."

Given this focus on exploring new territory, let me also acknowledge that there will be some simplifications made to complex problems, and some shortcuts taken in order to more clearly demonstrate a point. The purpose of this class is to teach the fundamental principles and relationships, not to get bogged down on the third decimal place - which doesn't mean that it's not important. I shall attempt to make clear those cases when I am purposely simplifying a complex issue, but I invite the reader to be alert to this and use her own wits to determine whether such a simplification would

Figure 2.2: The author's summer residence, SUNDANCE at anchor in Canada

be justified in any particular real-world design problem.

When we are exploring unknown territory we often need to acquire a new skill at a moment's notice: "How shall I ford this stream? How quickly can I teach myself the art of bridge-building?" In this light the AMV designer must be constantly in a learning mode, constantly acquiring new skills against the day when they may be needed.

And, since we are in territory not occupied by our conventional-ship brethren, we should expect to acquire skills and tools that are not in their toolboxes. Thus I say "Keep your eyes open: Look left, right, look outside your community. The idea you need may be behind you."

I hope in this course to introduce the student to the skills needed for AMV design. I hope to introduce you to the sources where some of those skills and tools are found. But, as I say to the undergraduate students, it has taken four years to teach you steel monohulls - I can't teach you fiberglass, titanium, and aluminum, catamarans, trimarans, SES, SWATH, and hovercraft in a single book.

3 The Search For Speed

The timeline of every ship design project proceeds something like this[1]:

- IDEA $\Rightarrow$

- INQUIRY $\Rightarrow$

- CONCEPT DESIGN $\Rightarrow$

- PRELIMINARY DESIGN $\Rightarrow$

- CONTRACT DESIGN $\Rightarrow$

- DETAIL DESIGN $\Rightarrow$

- CONSTRUCTION $\Rightarrow$

- THROUGH LIFE SUPPORT $\Rightarrow$

- BREAKING

The first *two* of these steps occur at the customer's facility (or in his mind). The first *three* are where the greatest whole-ship creativity takes place, and are often where the AMV solution first makes its appearance. These three are also, in the commercial world, called by the dirty word "sales". In this work I shall treat primarily of the Concept Design stage of the process, but I hold as axiomatic that good concept design is impossible without knowledge of detail design & construction, and ship operation/through life support.

The term "Advanced Marine Vehicles" or "AMVs" embraces a broad range of craft types. In most cases these vehicle types were invented in an attempt to attain higher speeds at sea than are possible with conventional ship types. Two exceptions to this rule are the SWATH - invented to gain exceptional ride quality - and the Hovercraft - invented to gain amphibious capability. As an overview, let us take a Cook's Tour of the world of AMVs. This tour will provide a brief introduction to the range of high speed hull forms that are currently in a naval architect's tool box.

---

[1]This assumes that the AMV design spiral is basically the same as that used for conventional ships. This may not be optimal, but it is often true.

At the beginning we are just going to pass the tools out and touch them. Imagine passing a box of tools to a group of children: "Johnny, this is called a 'hammer'..." And just like hand tools, yes it's true that you can drive a nail by hitting it with a screwdriver, or that you can use a claw hammer to turn a nut on a bolt, but that is not what each of those tools is optimized for. So we spend the first few pages just handling the tools in the toolbox, twisting and turning and looking at them from a number of different points of view, to learn what each of these tools of the trade - what each of these Advanced Hull Forms - is good for, optimized for, intended for, etc. Why do we do this? Because "to the man who only has a hammer, everything looks like a nail." The suite of AMVs represents tools (hull form choices) in the naval architect's tool box, and these tools allow him to undertake projects that may be impossible to the one-tool designer.

In brief, the message behind this presentation is that there are a variety of hull forms available, and that each has its own strengths and weaknesses, each has its own niche. There is no one hull form that is best for all applications, but instead it is helpful to understand what each concept brings to the table, and what each concept's limitations are.

3.1 What is Fast? What is Speed?

Since this text is about high speed ships, let me start with an introductory remark about speed and power.

What is Fast? Unfortunately, the answer depends upon size. A fast 100-foot boat may require quite a different hull form solution from a 1000-foot boat of the same speed. It is important to begin by understanding the relationship between speed and size.

Consider the vessel illustrated in Figure 3.1. Clearly this small speedboat is "fast." Equally clearly the rowboat in Figure 3.2 is "slow." But what about the ship shown in Figure 3.3? Is this vessel slow? In absolute terms this Washington State Ferry is faster than the rowboat. And it's probably faster than some Bayliners, no matter what their owners may claim in the marina bar. And yet despite its 20 knot speed, it is still in some sense "slow" and has more akin to the rowboat than it does to the speedboat. How can we resolve this conceptual difficulty? How can we recognize that speed seems to take on different meanings for big ships versus little ones?

The answer lies in the physics, and is captured for naval architects in the tool of "Froude Number." The Froude Number combines speed and size. In terms of Froude Number the kayak and the ferry are just about equal, while the JetSki's equal is found in the motoryacht DESTRIERO, depicted in Figure 3.4.

Figure 3.1: This small 20-knot speedboat is clearly "fast."

Froude Number is, crudely put, "speed divided by size." The "size" can be length, displacement, or many other things. Two particular formulations of Froude Number are the most common in naval architecture: "Length Based" and "Volumetric" Froude Number.

The Length-Based Froude Number is: $Fn_L = V/(gL)^{0.5}$ and is the most common in naval architecture. Volumetric Froude Number is useful in some high-speed ship problems, and is also used for regulatory purposes by IMO. The Volumetric Froude Number is: $Fn_{vol} = V/(g(Vol^{0.333})^{0.5})$, Vol is the volumetric displacement of the ship, in cubic meters or cubic feet.

The difference between these two Froude Number formulations may become important in some particular analyses, but these difference are unimportant to what we are talking about here. What we are talking about here is that Froude Number allows us to combine the effects of speed and size, so that when we talk about "fast" ships we mean either 20-knot 60-footers or 60-knot thousand-footers.

As an exercise, the reader is invited to find the displacement and speed of ten widely different vessels, and then calculate their volumetric Froude Numbers. Compare the results and see how that aligns with the concept of "equivalent speed given differing size."

Figure 3.2: This 2-knot rowboat is clearly "slow."

3.2 Higher Froude Number means More Power

We all know that higher speeds require more power, but looking at this truism through the lens of Froude Number can be particularly revealing.

Figure 3.5 presents a plot of the power and speed of a large number of vessels, and there is no pattern readily apparent. But we can apply some simple logic to bring order to this chaos:

Firstly, we know that a bigger (heavier) ship will require more power (for the same speed) than a lighter ship. So in Figure 3.6 I present the same data, but in this case the power has been replaced by the Specific Power, or Power To Weight Ratio. It's not much better than Figure 3.5. But in Figure 3.7 I replace the dimensional speed in knots, with a non-dimensional speed in Froude Number (in this case volumetric Froude Number.) Look at how much order this has imposed upon the data - there is a very clear trend revealed.

We will return to this type of analysis in a later chapter, but at this time I want to draw one simple conclusion: Going fast takes more power. Going to a higher Froude Number requires more power per tonne of ship weight.

Figure 3.3: Is this Washington State Ferry "Slow" or "Fast"? In numerical terms it is nearly the same speed as the speedboat, and yet in hydrodynamic terms it is as "slow" as the rowboat. This truth is captured through the naval architect's Froude Number.

Figure 3.4: Motoryacht DESTRIERO. Her Froude Number is approximately the same as that of the speedboat in Figure 3.1

The graph shows power required, per tonne of displacement, for a range of Froude Number. You can clearly see that as the ships go faster the power demand rises dramatically.

For the record, those Washington State Ferries that I called "slow" are at $Fn_vol = 0.90$, Specific Power $= 0.00045$; very slow, very low power. That 60-knot motor yacht is at Specific Power $=.035$ kW/kg, $Fn_vol = 2.55$. The only vessels out at the very high Froude Numbers, say greater than 5.0, are a handful of extreme craft.

Bottom line: Going fast takes more power, and "fast" is defined by Froude Number.

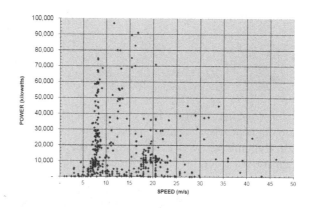

Figure 3.5: Speed and power data for a collection of vessels.

3.3 Hull Form vs Froude Number

There are different types of craft that are appropriate to different niches of the speed plane I have presented. Let us consider the primary choices of Advanced Marine Vehicle to be as follows (we will further define and describe each of these in following pages):

- Monohull
- Catamaran
- Hydrofoil

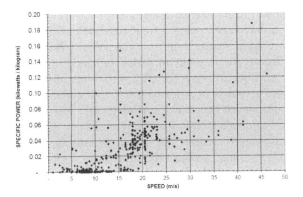

Figure 3.6: The same vessels as the preceding Figure, but now presenting Specific Power versus Speed

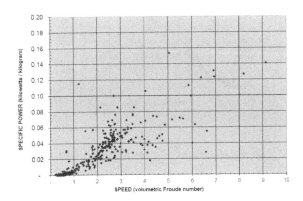

Figure 3.7: The same vessels as the previous two Figures, but now presenting Specific Power versus Non-Dimensional Speed (Froude Number)

- SES (Surface Effect Ship)

- ACV (Air Cushion Vehicle)

Figure 3.8 shows again the same data as the previous three figures, but herein the spots have been colored to show which of those five types each craft is. I have also "zoomed in" on the lower left hand corner of the graph to emphasize the domain in which lie ships of practical economic interest. Here you begin to see the niches for each of the hull types. The catamarans were invented in order to get speeds higher than the monohull range, and you may see that they appear to take less power than similar speed (Froude Number) monohulls. Surface Effect Ships and Hydrofoils were further invented to reduce the power demand at the highest speeds. Hovercraft (ACVs) appear to have the best speed performance. But before we go too far in this analysis, let's go look at some representative ships.

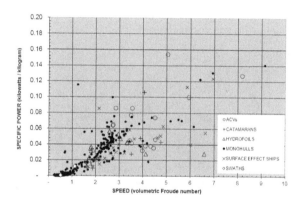

Figure 3.8: The same data as Figure 3.7, colored to show hull type

3.3.1 High Performance Monohulls

The most common hull type of course is the monohull. They have been around for millenia, they are extremely efficient versatile hulls. Some of the highest performance monohulls are painted grey, see Figure 3.9. The selection of the monohull form by these customers is not due to a lack of money, but is due to the extreme versatility and efficiency of this form, up to a volumetric Froude Number of about 1.

Figure 3.9: Most surface combatants are classed as High Performance Monohulls

To get to higher Froude Numbers we start running into a wall of resistance, both wavemaking and frictional. One technique for reducing resistance is to find some way to lift the hull, or a part of it, out of the water. This will reduce the wetted area that contributes to friction, and by reducing the volume of water that must be "shoved out of the way" as the ship passes it should also reduce the ship's wavemaking drag. One such hull is a planing hull. These craft deserve a course of their own, as the physics of planing lift contains some quite interesting phenomena. Planing craft will only be lightly touched upon in this present course, but they form an important baseline for the hull forms that follow.

There is some unscientific debate on the waterfront as to what constitutes planing with some empirical definitions being put forward that any vessel above some critical Froude Number "must be planing" or that "slender hulls can't plane" and so forth. These assertions are not true. The definition of planing is that some fraction of the craft's weight is borne by dynamic lift, regardless of any particular speed or hull feature. The test for planing is to see whether the craft is lifted by Bernoullian forces.

The planing hull form is commonly used for patrol boats and recreational craft (like the speedboat.) A planing hull is usually fairly blunt, with a length to beam ratio of around 3:1. Planing hulls are commercially employed on short-sea or coastal routes. Planing hulls yield service speeds up to about 50 knots (although smaller planing hulls do indeed exceed 100 knots.) Planing craft are generally small, say less than 40 meters, or less than a few hundred tonnes. (Again, there are exceptions to these generalities, such as the 60m/1000 tonne DESTRIERO already pictured. But it is her deviation from the norm that makes her worthy of picturing.)

DESTRIERO has already been illustrated in Figure 3.4. She is a private yacht, built to be the fastest ship to cross the Atlantic. She is 67m in length with a design displacement of 1000 tonnes. She attains speeds in excess of 60 knots, and has an unrefueled range of more than 3000 nautical miles, having crossed the Atlantic Ocean unrefueled in about 60 hours. She represents possibly the apotheosis of planing hull design, because she was the recipient of a nearly unlimited budget, with a very clear goal "to be the best." Her designer Mr. Donald Blount had spent a career in the design of military patrol craft, and brought a huge knowledge of planing hull design. He revelled in the DESTRIERO project, describing it once to me as "finally the chance to do every detail right."

Figure 3.10: The MDV3000 Fast Ferry JUPITER, built by Fincantieri

A commercial planing hull, built by the same shipyard that built DE-STRIERO, is the fast ferry JUPITER, depicted in Figure 3.10. DESTRIERO sails at an ambitious volumetric Froude Number of 2.5, but she carries very little payload. A real commercial payload-carrying ship is the Italian MDV-3000 monohull, which sails at a Froude Number of 2.0, or 44 knots, and can carry 1600 passengers and 250 cars.

3.3.2 Stabilized Monohulls

DESTRIERO and JUPITER are both planing monohulls. Planing is an attempt to make the ship go faster by lifting some portion of the hull out of the water - and it works. Another way to make a hull go faster is to make it extremely slender - using a very narrow beam. A narrow hull will produce lower pressures bow and stern, because she cleaves the water more gently. But when taken to extremes this results in an unstable ship, so some sort of outrigger has to be added to get stability. The result is the trimaran.

Figure 3.11: CABLE & WIRELESS ADVENTURER, built for the 1998 around-the-world record. (Photo from www.solarnavigator.net)

Trimarans belong to a class of vessel properly called Stabilized Mono-hulls. They are characterized by the extreme slenderness of the main hull, and the presence of some suite of stabilizing outrigger hulls. Note that while "trimaran" implies that there are three hulls total, there are in fact Stabilized Monohulls having one ("Very Slender Vessel") two (a "Proa") three (a trimaran) and five (the NGA "Pentamaran") hulls. All of these types fall into the class of Stabilized Monohull.

Figure 3.11 depicts the trimaran that held the record for fastest around the world trip, having completed an equatorial circumnavigation in less than 80 days in 1998. The picture clearly shows the extreme L:B ratio of the main hull, and the almost vestigial nature of the outriggers.

Figure 3.12 depicts the Austal Shipyards trimaran ferry BENCHIJIGUA EXPRESS , built in 2005 for Fred Olsen Lines for service in the Canary Islands. She is 127 meters long, with a displacement in the neighborhood of 3000 tonnes, a service speed of 40 knots, and a payload capacity of about 700 tonnes.

Figure 3.12: Photograph of the trimaran ferry BENCHIJIGUA EXPRESS, built by Austal Shipyards.

3.3.3 Catamarans

Slenderness allows designers to get speeds up to about Froude Number of 2. Slenderness can yield speed, but it introduces stability problems, and so the trimaran was invented. The same push to slenderness gives rise to the catamaran. The catamaran uses a very slender hull to get low drag, but it overcomes the stability problem by putting two of these hulls side by side. The gap between the two hulls is spanned by a "raft" structure, which is usually where the payload is carried. This results in a ship with lots of room, well suited for carriage of a high volume / low density cargo. And one example of such a cargo is: People. Catamarans make excellent ferries. For a denser cargo trade, such as, oh, say, oil tankers, we don't see any catamarans, because their spaciousness is not useful with such a dense payload, and indeed their somewhat more complex structure becomes a penalty, not a benefit. But for ferries they have fitted very well, and we have many impressive examples, some of which follow.

The first example, in Figure 3.13, is the Stena Lines HSS 1500 built in 1996, which is (I believe, as of this date) still the largest catamaran in the world. The pictures clearly show the twin-hull design, and the large box-like ferry deck that spans them.

Figure 3.13: The 122m Stena HSS 1500 catamaran ferry, in service on the Irish Sea.

A smaller catamaran ferry is depicted in Figure 3.14, the Washington State Ferry SNOHOMISH built in 1998, which may form an interesting contrast to the Monohull Washington State Ferry depicted in Figure 3.3. Of course, the car-carrying monohull and the passenger-only catamaran are not the same mission, and thus have very different characteristics - they merely share the same owner. But this highlights an important point: There isn't one right hull for all jobs - even a single owner may find it desirable to have a stable of different hull forms for different niches. As the English say: "Horses for courses."

3.3.4 Wave Piercing Catamarans

Catamarans have encountered some difficulties, and particularly in the early days there were some issues with ride quality. In an attempt to improve the ride, the Australian naval architect Phil Hercus invented the wave piercing hull form. This hull form concept equips each hull with a narrow protruding bow to pierce or knife through the waves rather than rising up over each one. The first of Hercus' wave piercing hulls in commercial service was the 1985 SPIRIT OF VICTORIA, shown in Figure 3.15.

Figure 3.14: The Washington State Ferry catamaran SNOHOMISH (now retired.)

Early experience with the wave piercing concept was promising, but when the waves got too big the problem was that the ship still pierced them, resulting in slams of the box-like passenger compartment. This was corrected by adding a central "third bow" that does not actually touch the water - See Figure 3.16.

All of the catamarans described up to this point are operating in the speed range of $FN_vol = 2.0$. There are cats that go faster, such as the one illustrated in Figure 3.17. This vessel operates in Argentina at a FN_vol of about 3.5. But to get up to these speeds we have to make some hull form changes. In particular, this boat, at about 50 knots, has now begun to marry the planing hull form with the catamaran.

3.3.5 Hydrofoil Assisted Catamarans

To further increase the speed of a catamaran above a Froude Number of 2, to, say, 3.0 or higher, some have tried to marry them to hydrofoils. As far as I can tell this was first proposed by Dale Calkins and Dr. Peter Payne (independently) in approximately 1977. Many years later, prototype craft were built in South Africa by E. G. Hoppe and Nigel Gee (again

Figure 3.15: Australian naval architect Phil Hercus' first commercial wave-piercing catamaran.

independently, in approximately 1990.) South African work continues today under Dr. Volker Bertram.

While this principle does work, there still aren't many real examples of Hydrofoil Catamarans on the water. Figure 3.18 shows one foil assisted cat that was built in Sweden a decade or two ago, and is no longer in service.

3.3.6 Hydrofoils

That brings us to traditional hydrofoils, which is to say monohull hydrofoils. These ships are, without doubt, the most comfortable, smoothest ride, of any of the fast ship concepts. Unfortunately they are also the most expensive by far. A fully-submerged hydrofoil will permit speeds up to Froude Number of 4 or higher.

Hydrofoil development was, like so much else, originally military driven. Figure 3.19 shows the USN hydrofoil patrol craft of which six were built in the period 1973 - 1982 (note that the foils are visible underwater in this photo.) They were all retired in 1993. These patrol craft were built by Boeing, who then developed the ferry product line known as the Boeing

Figure 3.16: The Theater Support Vessel SPEARHEAD, a military Wave Piercing catamaran, after the pattern invented by Phil Hercus.

Figure 3.17: The Argentine ferry PATRICIA OLIVIA II

Figure 3.18: A hydrofoil-assisted catamaran. Photo from www.foils.org

JetFoil, depicted in Figure 3.20. The JetFoil had a speed of 45 knots and carried 250 passengers in unparalleled ride quality. Acquisition of a hydrofoil is two- to three- times the price of a catamaran. The last data I had on Boeing JetFoils in 1995 was they were running 13 to 17 Million dollars at that time.

3.3.7 Surface Effect Ships

The next class of vessel are the air-cushion catamarans or Surface Effect Ships. These ships are also in the Froude Number 3-to-4 category. In this type of vessel a cushion of pressurized air between the catamaran-like sidehulls is used to lift the boat above the water. The result is an almost complete elimination of wetted surface, and thus low drag. The drawback is the mechanical complexity of the systems required to create and contain the air cushion. Here again we see a tradeoff between speed-power performance, versus other concerns such as simplicity and low cost.

Two alternative terms for an SES are "Sidewall Hovercraft" or "Air Cushion Catamaran." These two names are nice, because they capture the family relationship between an SES and its cousins, the catamaran and the hovercraft.

Figure 3.21 is one of the better looking (in my opinion) SES of the world,

Figure 3.19: US Navy "PHM" hydrofoil patrol craft

built in Norway in about 1991. She carries about 400 passengers with a 42 knot service speed. Figure 3.22 shows a Norwegian Navy patrol craft which is evolved from the earlier Cirrus work.

Of course, the landmark SES project was the US Navy program in the 1970s, and I can't resist showing just one or two pictures from those exciting days. The project was an R&D effort, and built two 80-foot test craft. The vessel shown in Figure 3.23 exceeded 100 mph.

3.3.8 ACVs or Hovercraft

Continuing on with the air cushion theme we come to the hovercraft. The ACV or Air Cushion Vehicle is a fully skirted craft, which does not have the catamaran side hulls of the SES, but is in fact more like an Air-Hockey puck. As a result of its total air cushion, it is an amphibious craft. It also has very low drag, permitting speeds higher than Froude Number = 4.

ACVs tend to be noisy, therefore a bit uncomfortable, and mechanically complex, but they do have unmistakably unique capabilities, such as the ability to fly up over the beach. Large hovercraft successfully served on the English Channel for over 25 years - see Figure 3.24. They have since been replaced by catamarans, since the route really didn't need their amphibious

Figure 3.20: A commercial Boeing JetFoil

Figure 3.21: Norwegian Cirrus 120P class Surface Effect Ship ferry, circa 1995.

capability.

I have seen some written materials which propose amphibious hovercraft

Figure 3.22: Norwegian Navy SKJOLD SES patrol craft, circa 2000

Figure 3.23: The US Navy SES 100B

for airport-to-airport service across San Francisco Bay.

Figure 3.24: The English SR.N-4 commercial hovercraft, which served across the English Channel for over 30 years.

3.3.9 Wing in Ground Effect or "WIGs"

Well, as long as we're flying above the ground, let's add the Wing in Ground Effect machine. There aren't any of these in commercial service, but they may have a niche, and they are a nice futuristic point to end on. Figure 3.25 depicts one of the ones that started it all, flying in the late 1970s. WIGs may have service speeds as high as Froude Number = 14, or more.

Figure 3.25: The Caspian Sea Monster - a Wing-in-Ground Effect (WIG).

4 The Sustention Space

Having now met the various types of advanced vehicles, it is easy to feel overwhelmed by their variety or diversity. So I like to begin our study by introducing a systematic taxonomy of vehicle types. Taxonomy is the science and practice of classification. We use taxonomies as a means of imposing order on what might otherwise appear to be an infinite cloud of choices and possibilities. By applying a taxonomic system we will find that the cloud falls naturally into clusters of related concepts and types, and that these clusters can be manipulated, studied, or understood, as families. Families share certain characteristics, so if we understand the families that are defined by the taxonomy, we find that we can reasonably anticipate the characteristics that any member of that family will have.

Why do I teach this systematic taxonomy? My reasons are:

- So you can identify any given AMV concept, by identifying that family of which it is a member.

- So you can guess what will be the strengths & weaknesses, or other special features, of a given AMV concept, because you know the strengths & weaknesses of its family in general

- Because this taxonomy is used in the community, and you want a common language with your peers

In this class we introduce two different taxonomies - the triangle and the cube.

4.1 The Sustention Triangle

The "sustention triangle" is a commonly used device for characterizing ship types. This triangle is illustrated below. It is a conceptual device for understanding what makes the boat float. Traditional ships float because they are immersed in water and buoyed up by Archimedes' force. This is called "buoyant lift" and occupies the lower left corner of the triangle.

There are other ways to hold ships up. The reader may be familiar with hovercraft, for example, where the ship is lifted on a bubble of air.

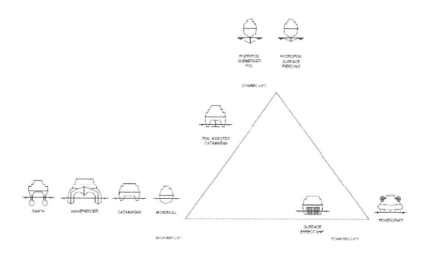

Figure 4.1: The Sustention Triangle, including illustrations of some of the
ship types at various points therein

Hovercraft have operated between England and France for thirty years now.
Hovercraft are examples of "powered lift" craft, as depicted on the lower
right corner of the triangle. Another lift type one may be familiar with is
"dynamic lift". A water ski works by dynamic lift. It does not float, but
when pulled fast enough through the water it generates a good lift force and
raises the entire payload up out of the water. Hydrofoils and hydroplanes
are both dynamic lift craft - the topmost corner of the triangle. Some craft
occupy intermediate positions on one or more edges of the triangle. For
example, an SES is part catamaran and part hovercraft. In fact the French
SES AGNES is part SWATH / part hovercraft.

4.1.1 The Problem With The Sustention Triangle

The sustention triangle is a good concept, has been in use for decades, and
has done good service. It does, however, have some flaws. In general these
flaws may be characterized by one typical example: The sustention triangle
is unable to distinguish between hydrofoils and WIGs: both are classed as
dynamic lift craft. Where then can we look for a sustention model that
does not suffer in this manner?

The sustention space concept attempts to provide a taxonomy of AMVs
according to the origin of their lift forces. Forces are generated by the
fluid(s) that a vehicle is passing through. Lift, by definition, is the com-

ponent of force perpendicular to the direction of travel. For designers of surface vehicles, our path being mostly horizontal, that means that lift is the net vertical force.

Lift is developed by the fluid pressures acting over the surface of the vehicle, in the water primarily, but also (for high-speed vehicles) in the air as well. Further, at zero speed, in calm water, the sum of all forces acting on the body had better be vertical; otherwise you could just set the vehicle down and it would take off in one direction or another. Moreover, the magnitude of that vertical force has to be equal to the weight, and the force has to act through the center of gravity, for equilibrium.

With speed, of course, the sum of fluid forces on the body surface can have horizontal components, which then become a part of drag (and in general, also, lateral forces that may be important in maneuvering).

For our sustention discussion we will look exclusively at the pressure forces on the vehicle. It is true that there are other forces, viscous forces, that act tangential rather than normal to the body. But unless there is a vertical component of the velocity, it is difficult to see how viscous forces can contribute much to supporting the weight of a vehicle. So we look primarily to pressures, and the integral of normal forces on the body, if we're interested in seeing where lift might come from

Where does fluid pressure come from? It has part that involves $\rho g h$, which we identify as buoyancy, and part that involves $1/2\rho v^2$, which is the so-called dynamic pressure.

Planing craft and hydrofoils are outside the scope of NAME4177, but that is only because they're well covered elsewhere. But we shouldn't let anyone conclude that "dynamic lift" doesn't happen (or even that it's small enough to consider unimportant) except on so-called "dynamically lifted" craft, that is, planing bottoms and hydrofoils. Dynamic lift is present to some degree on all high speed craft. Thus all types of craft have varying quantities of buoyant and dynamic lift. So it is perhaps best to think of "sustention space" in terms of parts of the total pressures that either do or don't involve v^2, rather than in terms of distinct breeds of craft (floating log vs. skipping stone).

In this conceptualization, air-cushion support is a wrinkle on buoyant lift at very low speed and a change in the boundary condition on the bottom (of the cushion, as compared with a hull bottom) at any speed. But the lift is still $\rho g h$, despite the fact that the "h" is created by fans.

Approaching the sustention question from this physics-based standpoint, we now are lead to a new model: The Sustention Cube.

4.2 The Sustention Cube

The Sustention Cube is a "design space" consisting of three mutually-orthogonal axes, as follows:

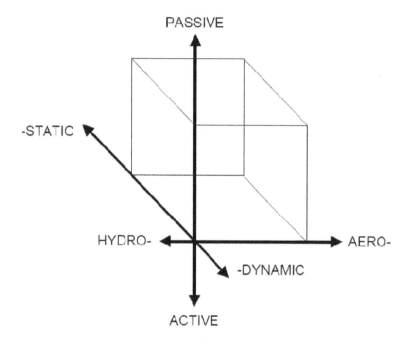

Figure 4.2: The Sustention Cube, the author's alternative model of the AMV design space. This model offers broader applicability by covering more of the design space than the Sustention Triangle.

4.2.1 First Axis: Static Lift or Dynamic Lift

Does the lift of the craft require that the craft be moving? Does the lift arise from $\rho g h$ or from ρv^2? The test for this is whether the craft's lift balance changes when forward speed is applied. Obviously planing craft change their lift balance as they come up to speed, thus clearly making them dynamic lift craft.

4.2.2 Second Axis: Aero- Lift or Hydro- Lift

Is the lift created by the displacement of air or of water? Barges are hydrostatically supported. Airships (blimps) are aerostatically supported. Hydrofoils and planing craft are hydrodynamically supported. Airplanes and WIGS are aerodynamically supported.

4.2.3 Third Axis: Powered or Passive

Alternatively these terms may be "active" or "mechanical" versus "passive." The test for this is whether the lift is due to the active motion of some component of the craft, or on the other hand is the lift due to the basic shape (geometry) of the craft? Most ships get their (static) support from their hull form, thus making them passive hydrostatic craft. Note that planing craft and airplanes should be labeled as passive dynamic craft. They require power to generate the speed that activates their lift, but the lift itself is the result of the shape of the bottom, or the shape of the wing.

This definition is the hardest to grasp of all those present in the Sustention Cube taxonomy. In particular it seems difficult for some people to grasp the distinction between "Passive/Active" and "Static/Dynamic". As one attempt to clarify this, I offer the following:

Does it depend on motion, or does it depend on switching on some component of the ship? A passive dynamic craft will generate lift if you simply tow it up to speed, whereas an active craft such as a hovercraft will not hover if its fans are off, no matter how hard you tow it.

Thus my test for "Active vs Passive" is: Can it be switched off, independent of the propulsion of the vehicle?

The clearest example I have is a Hovercraft: A hovercraft floats on a bubble of displaced water. It is perfectly happy to float "on cushion" at zero speed. In this case there is a volume of water (equal to the weight of the vehicle) which is displaced by the pressure of the air. And, clearly, if the fan is switched off and the air pressure escapes, the Hovercraft will cease to float.

By contrast a planing boat definitely requires ahead speed to plane, but it does not "care" how this speed is produced: It may be self propelled, it may be blown forward by a hurricane-strength wind, or it may be towed on a rope.

I acknowledge the conceptual difficulty in distinguishing between these, and I do not mean to belittle those who have this difficulty. Instead, I hope that the above discussion offers some small improvement in the comprehension.

4.3 The Corners of the Sustention Cube

The last description above now leads us into discussions of the total shape of the sustention cube, which may be defined by labeling the corners. The corners are defined by combing the following pairs, to produce eight points:

- Passive or Active

- Hydro- or Aero-

- -Static or -Dynamic

Thus the eight corners are:

- Passive Hydrostatics

- Passive Hydrodynamics

- Passive Aerostatics

- Passive Aerodynamics

- Active Hydrostatics

- Active Hydrodynamics

- Active Aerostatics

- Active Aerodynamics

Let us now consider the population of ships that lie inside this cubical space.

5 The Contents of the Sustention Cube

The Sustention Cube provides us with a taxonomic system for differentiating the AMVs according to features of their sustention. What we will find throughout this course is that their sustention also dictates some features of their performance - that, for example, all aerodynamic vehicles have generally similar performance, as contrasted with their, say, hydrostatic cousins.

5.1 "Fast, Comfortable, and Cheap: Pick any two."

In order to see this more clearly, it is helpful to define a set of performance parameters which allows us to track the performance of these vehicles. While in my title I suggested a three-parameter performance space, I actually prefer to use a five-parameter space as follows:

- Seakindliness

- Speed/Power

- Comfort & Space

- Load Carrying Ability

- Economics (Acquisition & Operation)

The naval architectural challenge is to balance competing requirements or desires, so I will apply subjective judgement to assess the relative merits of each of several AMV types. I publish these assessments with some trepidation, because they are in fact subjective. But I offer them as providing context for the discussion that follows.

In terms of the five "dimensions" of the performance space, I evaluate the principal types of AMV as shown in Table 5.1. (In the table, a higher number is better.)

Hull Form	Sea-kindliness	Speed / Power	Comfort & Space	Load carrying ability	Economics
Catamaran	2	1	3	2	3
Trimaran	2	2	2	3	3
SWATH	3	1	3	1	2
Hydrofoil	3	3	1	1	1
WIG	3	3	1	1	1
ACV	1	3	3	2	1
SES	2	3	3	2	1

Table 5.1: The author's subjective assessment of various AMV hull forms against five performance parameters. A high number indicates better performance.

A good discussion of the quest for speed at sea, and the various types of AMVs that have resulted, is presented by Clark et al, Reference [11], available online. The authors also present useful comparisons of the capabilities of the AMVs in the other performance areas such as seakindliness, etc. Note also that Speed is addressed in terms of speed in a seaway, and not merely speed in calm water. The degree to which wave conditions are expected will change the degree to which one hull type is preferred over another...as the authors discuss. Permit me to now marry the sustention taxonomy with the five parameter performance space, and let's see if we can't begin to recognize some patterns in the universe of AMVs.

5.2 A Note On Scalability

In the paragraphs that follow I refer to the scalability of the various AMVs in terms of their adherence (or not) to the cube/square law. In teaching this course to undergraduates I find that not all of them are familiar with the cube/square law, so I insert here an explanation derived from the Wikipedia entry on this subject:

The **square-cube law** (or **cube-square law**) is a principle, drawn from the mathematics of proportion, that is applied in engineering and biomechanics. It was first demonstrated in 1638 in Galileo's *Two New Sciences*. It states:

"When an object undergoes a proportional increase in size, its new volume is proportional to the cube of the multiplier and its new surface area is proportional to the square of the multiplier."

For example, a cube with a side length of 1 meter has a surface area of 6 m^2 and a volume of 1 m^3. If its side length were doubled, its surface area would be increased to 24 m^2 and its volume would be increased to 8 m^3. This principle applies to all solids.

In engineering, when a physical object maintains the same density and is scaled up, its mass is increased by the cube of the multiplier while its surface area only increases by the square of said multiplier.

Let us now consider that this object is subjected to dynamic pressures: Perhaps it is an airplane, flying because of the lift pressure on its wings. Perhaps it is a planing hull, skimming on the sea surface because of the dynamic pressure on its hull. The cube-square law means that if we double the size of the craft, we will get eight times the weight, but only four times the wing area (or planing area.) If the object is still going to fly, then carrying 8x weight on 4x wing means that the wing pressure must be doubled.

For a dynamic surface to generate double the pressure, we must either make a breakthrough in its lift pressure coefficient, or we must increase its speed, since the lift is defined as $L = C_L \frac{1}{2}\rho S v^2$ A breakthrough in lift coefficient is unlikely, so we resort to driving the vehicle fast = 41 percent faster in this case. [1]

So there I am: I have designed a one-man airplane that takes off at 50 knots. Encouraged by my success I decide to build one that carries eight people (eight times the weight.) And I am disturbed to learn that I must drive this new plane much faster - 70 knots - just to take off.

No problem, let's put in a bigger engine. We assume that the resistance of the airplane is $R = C_R \frac{1}{2}\rho S v^2$. We know that S went up by a factor of four, and we know that v^2 went up by a factor of two, so we expect that resistance will go up by a factor of eight. No problem, we need eight times the power, and that's an engine that weighs eight times as much, and that's consistent with all the stretching and scaling that we have already done.

Wrong! We need eight times the *thrust*, but recall that power is thrust-times-speed ($P = R * V$). So really the power doesn't go up by 8:1, it goes up by $8 * 1.41 = 11.3 : 1$. So now I have to find an engine with eleven times the power, not eight times. And it weighs 11 times as much. And that eats up my payload capacity so now I can't carry the eight passengers I want. So I make the whole airplane bigger still... And the spiral never converges.

The scalability of my airplane has been killed by the cube-square law, that forces increase as the cube of lambda, but areas increase as the square.

---

[1] 41 percent because the square root of 2 is 1.41

Effectively, if the physics of the vehicle is cube/square sensitive, then the vehicle is *not* indefinitely scalable.

5.3 The Advanced Marine Vehicles

In the previous sections we discovered that there are a range of vehicle types, each being generally suited to a particular speed niche. We then introduced a taxonomic scheme for characterizing these vehicles. Let us now employ that taxonomic scheme for taking a second walk through the AMV design space, focusing this time on understanding why we might choose one of these types over another, and what design challenges our choice will engender. Note that this tour will follow the eight vertices of the Sustention Cube, videlicet:

- Passive Hydrostatics

- Passive Hydrodynamics

- Passive Aerostatics

- Passive Aerodynamics

- Active Hydrostatics

- Active Hydrodynamics

- Active Aerostatics

- Active Aerodynamics

For each of the occupants of these corners, I will attempt to characterize their performance, in broad terms, in the five performance parameters of:

- Seakindliness

- Speed/Power

- Comfort & Space

- Load Carrying Ability

- Economics (Acquisition & Operation)

I leave it as an exercise to the reader to see if there might not be some graphical representation of this mapping.

5.3.1 Passive Hydro Static (Buoyant) AMVs

Conventional ships and barges. 20+/- courses in the study of naval architecture, and yet it's only one of the eight vertices of the sustention cube. Now in one course we are going to address not only this one corner, but also the seven others.

Buoyant craft include the majority of the ships in the world, including the high-performance monohulls I mentioned earlier. But in the context of Advanced Marine Vehicles the most important buoyantly-supported craft are the Multihulls and SWATHS.

Multihulls

"Multihull" of course means a ship with more than one hull. In conventional parlance this generally means displacement catamarans and trimarans - we don't usually refer to SWATHs and SES as "multihulls", although they are. Displacement multihulls owe their origin to certain observed facts about buoyant hull design - by this I mean that displacement multihulls are in fact derived from displacement monohulls. Displacement monohull design is "classic" in naval architecture, and is very well understood. Monohulls represent the most versatile hull form choice. However, as is well known, the monohull form gets into a bind when you try to make it go fast. In order to reduce drag for high speed, the designer is pushed to make the hull as slender as possible, thus reducing both pressure and form drag. The problem is that a slender monohull is difficult to make stable. How to make a skinny hull stable? Answer: Tie two or more of them together.

Catamarans The Wikipedia has a good general article on catamarans at: http://en.wikipedia.org/wiki/Catamaran. The word "catamaran" is derived from a Polynesian word meaning "multiple logs tied together" or in other words a multihull. In current usage a catamaran has specifically two hulls, generally identical.

The defining feature of the catamaran is both its two-hulled nature and the slenderness ($^\sim$20:1 L:B) of those hulls. The sustention of a catamaran is Buoyant or Passive Hydrostatic.

A pioneer of commercial catamarans was the Australian firm "INCAT" - short for International Catamarans. INCAT developed the variant of the catamaran called a wave-piercing catamaran, depicted in the previous chapter. Of course, just to keep you on your toes there are two firms named INCAT: A shipbuilding firm and a design firm. INCAT the shipbuilder is still in operation, and their website is http://www.incat.com.au/ INCAT the design firm is no longer in business under that name: The intellectual property of INCAT Designs - Sydney, Pty Ltd was sold to two firms. Data relevant to vessels over 60m in length was sold to Alion Science and Technology of the USA. Data relevant to vessels 60m in length and under was sold to Crowther Multihulls, who re-branded under the name "INCAT-Crowther."

Seakindliness: Neither a strength nor a weakness. The ship is buoyantly supported, so her seakeeping is buoyancy-dominated and subject to the same physics as a displacement monohull. There is a design challenge in that GM_T and GM_L tend to be similar, leading to corkscrew motions. In addition GM_T is high leading to snap roll. Cross structure can slam. Bow diving can occur in following seas.

Speed/Power: A strength of the catamaran: Slender hulls give good speed-power characteristics by reducing the wavemaking resistance.

Comfort & Space: In addition to being desirable for high-speed hydrodynamics, cats are also sought in low-speed applications where a lot of arrangeable area is needed at very low density. Arrangeable area is large per tonne of displacement. (A mental model that I use when understanding this is to imagine a bow view of a catamaran and realize that "there's nothing supporting the middle of the ship.") As a consequence this ship type is suited to low-density payloads or missions, such as the carriage of people.

Load Carrying Ability: See above. Also note that large arrangeable area can be a weakness in some applications (e.g. warships.)

Economics (Acquisition & Operation): Generally good. Lightweight construction is needed which causes some increased cost (compared to a steel monohull) but reduced powerplant size offsets this. Other ship systems are generally conventional so costs are also conventional.

Alternate Configurations: SWATH, Semi-SWATH, Wave-Piercing, and Foil Assisted

Nomenclature and terminology:

- Hulls (NOT "pontoons")

- Wet Deck (term derived from SES parlance)

- Tunnel

- Z-Bow or WavePiercing hull

- Third Bow (option, usually only found on wavepiercers)

Scalability: Unlimited (cube/cube)

Figure 5.1: The first of the INCAT 74m Wave Piercing Catamarans - HOVERSPEED GREAT BRITAIN, who then held the record for the TransAtlantic Crossing.

Trimarans A catamaran is an attempt to make a very slender hull, and give it stability by using two identical hulls side by side. The trimaran - properly called a "stabilized monohull" - is a similar attempt to make a hull very slender but give it stability by using one or more very small outrigger hulls. These outrigger hulls are usually made to be as small as possible, so as to minimize their resistance and structural penalties, while still being big enough to yield the required stability for the main hull.

A rather exotic looking trimaran is depicted in Figure 5.2.

Defining Feature: By definition, three hulls. But actually this term may be applied to any outrigger-stabilized monohull. The main hull is slender, say 20:1 L:B.

Sustention: Passive Hydro Static (Buoyant)

History: Trimarans are of ancient origin, dating at least to native craft of pre-history. Modern interest in trimarans has grown slowly from early work in recreational craft, reaching the current peak in activity lead by Australian shipyard Austal, who have developed the 127m trimaran ferry BENCHIJIGUA EXPRESS and the related US Navy warship the LCS - See Figure 5.3 & Figure 5.4

Seakindliness: Long for its displacement yields good seakeeping. Buoyancy-dominated physics, as with any hydrostatic craft.

Speed/Power: Very high slenderness yields good speed/power character-

Figure 5.2: The Earth-Race trimaran, the most exotic looking trimaran I have come across

istics. Optimization of the amas (outrigger hulls) is tricky.

Comfort & Space: Generally somewhere between monohull and catamaran in arrangeability. Slender hulls may be difficult to fit machinery into.

Load Carrying Ability: Generally somewhere between monohull and catamaran. (There is "something holding up" most of the ship, except under the wings which reach out to the amas.)

Economics (Acquisition & Operation): Generally good. Lightweight construction is needed which causes some increased cost (compared to a steel monohull) but reduced powerplant size offsets this. Other ship systems are generally conventional so costs are also conventional.

Alternative Configurations: Pentamaran, Proa.

Nomenclature and Terminology: The outrigger hulls are called "amas" although this term is not well known outside the trimaran community.

Figure 5.3: The Austal trimaran ferry BENCHIJIGUA EXPRESS. Photos from www.austal.com

There is no accepted term for the cross-structure which connects the amas to the main hull. I prefer the term "wing" for this.

Other important terms are the "separation" referring to the distance that the amas are athwartships from the main hull, and the "stagger", which refers to the relative fore-and-aft location of the amas compared to the main hull.

Scalability: Unlimited (cube/cube)

SWATH - Small Waterplane Area Twin Hull   The SWATH is a type of catamaran designed specifically for minimum motions or maximum seakindliness. SWATH is an acronym for "Small Waterplane Area Twin Hull." It was coined, I believe, by Dr. Colen Kennell in the 1970s.

Defining Feature: The defining feature of the SWATH is the small waterplane area it possesses. This is usually manifest in a pair of torpedo-like lower hulls which are positioned some depth below the free surface by a set of surface-piercing struts. A SWATH may have one or two struts per side, and it is not clear how thick the struts can be before the SWATH ceases to be "small waterplane area" and becomes simply a catamaran. Indeed, some catamarans attempt to improve their ride quality by adopting small waterplane area in the forebody and calling themselves "semi-SWATH"

Figure 5.4: Austal's US Navy Littoral Combat Ship "LCS 2" in drydock

designs.

The best single-volume treatment of SWATHs is the SNAME T&R Bulletin "SWATH Ships" Reference [12]

SWATHs made a transition into 'mainstream' naval architecture when the US Navy built two classes of SWATH Ocean Surveillance ships, the T-AGOS 19 & T-AGOS 23 class. Figure 5.6 through 5.8 depict the T-AGOS 19. Another notable USN SWATH was the stealth ship "Sea Shadow", not pictured. Since these Navy projects, SWATHS have shown up in many other conventional naval architecture portfolios, such as the German pilot vessel marketed by Abeking & Rasmussen shipyard - See Figure 5.9.

Sustention: The sustention of a SWATH is Buoyant or Passive Hydrostatic.

History: A brief history of SWATH development, including some important progenitors that did not use the SWATH name, is found at: http://www.sw Several photos are found at: http://www.geocities.com/dthigdon/dynamics/im Don Higdon (the owner of that website) was instrumental in the design of the ride control systems for several of those vessels.

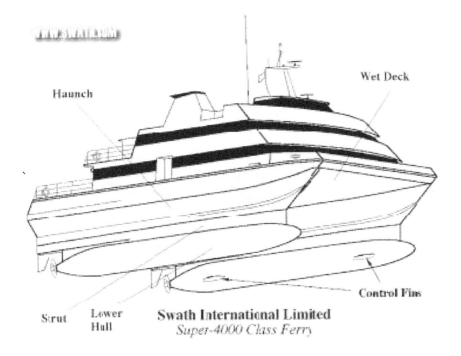

Figure 5.5: The parts and nomenclature of a SWATH. Picture taken from www.swath.com

Seakindliness: The advantage of a SWATH is that it is relatively decoupled from the excitation forces caused by surface wave action. This is accomplished as a direct result of the Small Waterplane Area.

Speed/Power: Low wavemaking resistance possible (not assured)

Comfort & Space: Catamaran-like

Load Carrying Ability Generally catamaran-like, except that the low waterplane area means a large change in draft or trim with load condition. Usually a ballast system is fitted to aid in maintaining desired attitude.

Economics (Acquisition & Operation): Good - Conventional ship technology.

Nomenclature and Terminology: The SWATH geometry has its own nomenclature, as follows:

- Hulls or "Lower Hulls" (but NOT "Pontoons")

- Struts

Figure 5.6: US Navy T-AGOS 19

- Wet Deck

- Haunch

- Controls Fins, consisting of "Canards" forward and "Stabilizers" aft

SWATHs also present some definition questions, the most important one being what is the length? In order to be unambiguous, we early decided that the definitive length should be the length of the submerged hull. This way it wouldn't depend on whether we were talking about a single-strut (per side) or a two-strut design.

Scalability Unlimited (cube / cube.) But the advantages vanish when ship size becomes very large.

Challenges: High wetted surface means generally not a high-speed hull form. Maneuverability challenges. Large beam and draft (may have shiphandling / docking challenges.) Submerged protuberances. Small waterplane area makes it weight / trim sensitive.

Alternate Configurations:

- SLICE- a four-legged variant. See Figure 5.10

Figure 5.7: US Navy T-AGOS 19

- Lifting Body Ships - Variants in which the submerged buoyancy (the lower hulls in a conventional SWATH) are merged into various blended shapes.

5.3.2 Passive Aero Static (Air Buoyant) AMVs

These craft exist: They are Blimps, Zeppelins, hot air balloons, etc. As airships they do have important roles to play in maritime affairs, and historically it is interesting to note that at the turn of the 19/20 century they fell within the domain of the naval architect, since they were Archimedean in support and dominated by so many of the same engineering concerns as "wet" ships. However, notwithstanding that interesting historical note, they lie outside the domain determined for this course.

5.3.3 Passive Hydro Dynamic (Dynamic Lift) AMVs

Dynamic lift craft get their lift from speed. When they stop, they sink. (Or they transform into some other kind of craft.) A man on a water ski

Figure 5.8: US Navy T-AGOS 19

is perhaps the 'classic' example of a Dynamically Supported Craft. At rest he is fully immersed, but above some critical take-off speed he becomes a flying machine. In the realm of Advanced Marine Vehicles the two that "really matter' are the hydrofoils and planing hulls.

Planing Craft

Planing craft are deserving of a course unto themselves, and indeed in most institutions (including UNO) they receive one. As such I have not attempted to include them in the AMV course. It can be argued that this is because this course deals with novel or unusual craft, craft for whom there is not a large body of experience and thus for whom the skills of Lewis & Clark are needed. This is not the case with planing craft, which have been studied in detail for at least half a century. Thus my choice to excluding them from this course is not a statement of their unimportance, but rather a statement of their relative maturity and thoroughness of treatment elsewhere.

Hydrofoils

Once class of dynamically supported vehicles is however not included in planing craft design courses, and that is the hydrofoil.

Figure 5.9: SWATH Pilot Vessel from German shipyard Abeking and
Rasmussen

A hydrofoil is a vehicle supported on wing-like structures immersed in
the water. The lift generated by these water-wings lifts the hull of the ship,
thus reducing the drag of that hull.

Defining Feature: The defining feature of the hydrofoil is thus the presence of the foils themselves - wing-shaped lifting surfaces. If these wings
are present, and they lift a substantial fraction of the craft's weight under the design condition, then the craft is a hydrofoil. Excellent resources
on hydrofoils may be gleaned by perusing the website and archives of the
International Hydrofoil Society, www.foils.org

Sustention: Passive Hydro Dynamics. The lift is caused by hydrodynamics (moving water forces), but this lift is generated passively, requiring only
the forward motion of the craft.

History: Hydrofoils have a remarkably long history, indeed, Alexander
Graham Bell experimented with hydrofoil craft as early as 1911. For some
enchanting histories of hydrofoils, see the following websites:

- http://www.histarmar.com.ar/InfGral/Hidroalasbase.htm

Figure 5.10: Four-hulled SWATH variant SLICE

- http://www.lesliefield.com/other_history/alexander_graham_bell_and_the_hydrofoils.htm

- http://www.foils.org/popmags.htm

- http://www.foils.org/pioneers.htm

Seakindliness: Hydrofoil craft of the Fully Submerged type (see below) are very well isolated from sea surface excitations and thus may have excellent seakindliness. In ferry service hydrofoils are well known to be the smoothest ride available.

Speed/Power: The hydrofoil itself produces a drag due to lift, and a drag due to the wetted surface of the foil. But these forces are much smaller than would be the drag of the hull if fully immersed and travelling at the same speed. As a consequence, a hydrofoil can attain substantially higher speeds for a given thrust than can a competing buoyant type craft.

The challenge with this is that the foil lift depends upon speed squared, (unless the foil C_L is modified), this means that the weight borne by the foil likewise varies as speed squared. In other words a fairly small variation in speed can cause a substantial change in the amount of reliance that is placed upon hull buoyancy, and thus the amount of hull drag introduced.

In consequence a hydrofoil is usually optimal only across a quite narrow band of operating speeds.

Comfort & Space: Hydrofoils are generally monohull-based, and thus have monohull-like arrangeability and space. There are some instances of catamaran-based hydrofoils. Also, in the case of the Boeing JetFoil one may note that the designers took hold of the vestigial or secondary role of the buoyant hull and made a quite unusual monohull, having more space than might otherwise have been given. Thus there is considerable flexibility available.

Load Carrying Ability: The load carrying ability of the hydrofoil is again generally monohull-like, always considering the fact that the lift varies as speed squared. This also affects the scalability of the craft, see below.

Economics (Acquisition & Operation): Hydrofoils are quite expensive. First, the foils are challenging to manufacture, demanding close tolerances and expensive materials. Note that a foil that lifts a 500 ton craft is a thin plank that is carrying a weight of 500 tons. It is common to find that designers resort to titanium or other high-strength materials for the construction of the foils.

In addition, Note that with a fully-submerged foil, there is no change in the lift when the craft encounters a wave. This is good for the ride quality, but in high seas it means that there is no "natural" wave-climbing behaviour. Indeed, there is no physics that tends to maintain the craft at a given fly height even in calm water - it all must be accomplished by a flight control suite (usually called Ride Control.)

Alternative Configurations: Configuration alternatives commonly encountered in Hydrofoils are as follows:

Hull type: Monohull or catamaran

Foil Submergence: A "Surface Piercing" hydrofoil has foils that penetrate the sea surface, see Figure 5.11. This configuration means that as they encounter waves they will generate additional lift and help raise the craft above the waves. They will also rise as speed increases, meaning that the foil lift coefficient can be maintained more or less constant as the craft accelerates.

By contrast, the fully-submerged hydrofoil has "wings" that are below the sea surface - see Figure 5.12. This results in a very smooth ride, but it requires a flight control system to balance the craft and to manage wave encounters.

A third category might be argued, which is "foil assisted" craft wherein the foils do not lift 100% of the craft weight, but only some lesser fraction. Properly these might be considered to be hybrid craft who sit along an edge of the sustention cube, rather than at one of its corners.

"Canard" versus "Airplane" configuration: The second major configura-

Figure 5.11: A Surface-Piercing hydrofoil produced by Rodriquez

Figure 5.12: A hydrofoil craft having fully-submerged foils. (The foils are visible below the sea surface in this photo)

tion choice concerns which of the craft's foils carries most of the weight. In the "Canard" configuration the forward foil carries most of the weight. Figure 5.11 is a canard configured craft. In the Airplane configuration most of the weight is carried on the aft foil, as in the case of the craft in Figure 5.12.

(Do not be misled by the choice of these two figures to illustrate this point - there is no necessary relationship between the choice of surface-piercing versus fully-submerged, and the choice of canard versus airplane.)

Scalability: Limited, perhaps to ¯1000 tonnes due to cube / square relationship.

The strength of the hydrofoil is its excellent speed / power characteristics, and excellent seakeeping for fully-submerged types. Their weaknesses are the narrow economic speed range, and the expense.

5.3.4 Passive Aero Dynamic (Dynamic Lift) AMVs

Passive Aero-Dynamic Craft are vehicles that require ahead-speed to fly (-dynamics) and their lift is generated by air, not water. This includes airplanes, which are clearly outside the domain of this course. But it has been decided that Wing-in-Ground Effect (WIG) vehicles are ships, and thus they will be touched upon here.

WIGs

A WIG is a wing which flies very close to the surface (either sea or ground) in order to benefit from the image system that appears in such case. (A full discussion of the image system is outside the scope of this course.) By exploiting the image system the lift-to-drag efficiency of the wing is much improved, resulting in very impressive craft performance.

A WIG attains this efficiency by operating within about one wing-chord of the surface. Above this height the benefit due to the image system falls off rapidly.

WIG's were invented, well, they were invented by God - see Figure 5.13. But they have been commercially developed in both Germany and Russia. Figure 5.14 shows one of the most impressive of the Russian military WIGs, the Caspian Sea Monster.

Rozhdestvensky [13] provides Figure 5.15 which depicts the increase in Lift to drag ratio (L/D), which for our purposes is equivalent to saying "the reduction of drag" that is produced when a wing operates in ground effect. As may be seen it is possible to double the lift of a wing (or halve the drag) by operating in extreme ground effect - say 5 percent of the chord length above the surface.

Real WIGs do not operate in so extreme a condition, but appear to be in the range of ten percent chord above the surface, at which height the benefit is a still-respectable 20-percent reduction in drag.

Figure 5.13: The Prototypical Wing In Ground Effect

Figure 5.14: The Caspian Sea Monster

Defining Feature: The defining feature of a Wing In Ground Effect is the wing, and its proximity to the ground. The key feature is to determine that this craft is aerodynamically supported. One does sometimes encounter WIGs which also incorporate air cushions or other features (usually as take-off and landing aids.)

Sustention: Aerodynamic, passively generated by the shape of the wing.

History: WIGs, as marine vehicles, are of fairly recent generation, say

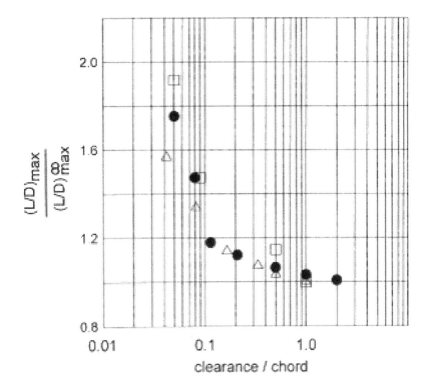

Figure 5.15: Rozhdestvensky's illustration of the benefit of ground effect upon vehicle Lift/Drag ratio

within the past 50 years. Pioneers in this field include Jörg, Lippisch, and unnamed scientists in the Soviet Union.

Seakindliness: A WIG flies approximately one wing-chord above the mean sea surface. If this chord length is large enough, then this can mean a height substantially above the waves in that surface. This means that a WIG can be nicely isolated from the roughness of the sea, yielding a very good ride quality.

Speed/Power: WIGs are fast, like airplanes. WIG speeds may be on the order of several hundred knots.

Comfort & Space: WIGs suffer from being airplane like in configuration, with that "mailing tube" shape which impairs their ability to transport bulky cargo. WIGs have been used as personnel transports. I know of no instances of WIGs carrying inanimate cargoes.

Load Carrying Ability: I don't know. As an aerodynamic vehicle I assume that they have a carrying capacity generally like that of an airplane, and I have seen airplanes of quite large capacity. What the limits are in this regard, and how these ratios compare to those of hydro- supported craft I don't know.

Economics (Acquisition & Operation): They have control systems and components like airplanes, and I suspect that they cost like airplanes. However it is worth noting that the Soviet WIGs (such as the Caspian Sea Monster) were built in shipyards, not in airplane factories. In view of this I hazard a guess that WIGs are somewhere intermediate in cost between ships and aircraft.

Alternative Configurations: There are many variants of WIG, including some of the more extreme of Tunnel Boats today. Figure 5.16 depicts a tunnel boat that is, in fact, a WIG.

Lippisch built WIGs of reverse-delta configuration, see Figure 5.17. Jörg on the other hand preferred a tandem wing configuration as Figure 5.18.

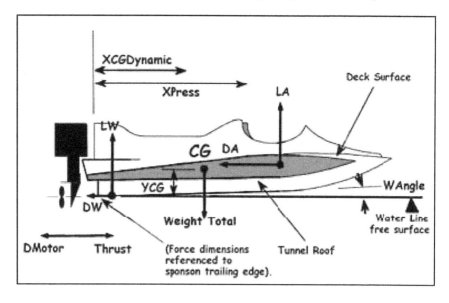

Figure 5.16: This illustration of the forces on a tunnel boat (from www.screamandfly.com) highlights the fact that these craft too are WIGs

Nomenclature and Terminology: WIGs fly by operating in a strong aerodynamic image system. This gives rise to the important Russian word

Figure 5.17: The Reverse-Delta configuration preferred by Anton Lippisch

Figure 5.18: A tandem-wing WIG craft from Gunther Jörg

"Ekranoplan" or "Screen Plane." The word "Screen" refers to the mirror image that yields the WIG's efficiency.

Note also in Figure 5.14 the very large tail surface. This tail flies out of ground effect itself, and is essential to providing pitch stability for WIGs. In fact, the frequent blow-over accidents of tunnel boats are due to the fact that they don't have these tail surfaces (because their designers don't know that they are actually designing WIGs.)

The WIG can sometimes be hard to take off, since the wing's lift develops as speed squared it takes some substantial speed before the wing is lifting the craft. To overcome this designers incorporate various take-off aids. In the case of the Caspian Sea Monster note the eight large turbofans, of which only two are needed for cruise flight. The other six are fired up only for take-off.

Scalability: The WIG is a dynamic craft and thus subject to cube/square limits. The upper limit in practical WIG size may be in the neighborhood of 1000 tonnes - although I am guessing at that figure.

The WIG has the lowest resistance of any AMV, and excellent tolerance for waves. It's weakness is that it is a little too much like an airplane, and many regulators don't quite know how to handle it: Does it require a pilot's license or a Captain's license? There are also challenges associated with maneuvering WIGs (they can't bank very far, so the turns must be flat slides). There are certainly challenges in docking and drydocking craft of this shape.

5.3.5 Active Hydro Static (Powered Lift) AMVs

Hydrostatic displacement means that the craft displaces a volume of water equal to it's weight. This is usually accomplished by pushing that water out of the way with some sort of impermeable structure, whether it be steel plates or rubber membranes. But anyone who has washed dishes in a sink knows that this same displacement can be accomplished by using an air bubble, such as in a bowl or drink glass turned upside down. The bowl will displace a volume of water and may float - although it is probably unstable.

It may be more surprising to realize that the glass need not retain the air bubble passively - the air bubble may be created actively. We can imagine some Rube Goldberg contraption involving a Shop Vac and a colander, which would end up floating just as well as the bowl first referred to. Indeed, the principal of this sort of sustention gives rise to a very important class of marine vehicles, which we know as hovercraft. They occupy the "Active Hydrostatic" niche of the sustention space.

A drinking glass upside down in the bathtub will float due to the bubble of air captured in the glass. This is passive hydrostatics. Now imagine the drinking glass with a hole in its bottom, and a fan to keep the air from getting out. This is Active Hydrostatics, and is exactly how a hovercraft works: It is supported by Archimedes' principle in water, but the displacement is created by fans.

ACV - Air Cushion Vehicle (hovercraft)

Using an air cushion as illustrated above will eliminate friction. Craft that employ this means still float by displacing water, it's just that they displace water due to the use of a machine (a fan): Active Hydrostatics.

Maki et al (2013) describe this well. They are writing about SESs (see below), but the description is an apt explanation of ACVs as well:

> *The Surface-Effect Ship (SES) is an attractive concept for applications that require a vessel to travel at high speeds. The vertical force that balances the weight of the craft is generated by a combination of buoyancy, hydrodynamic lift, and air-cushion support. For very high speeds, it is known that the wave resistance is small and the frictional drag dominates the total resistance of the vessel. The principal mechanism in which the SES shows an advantage over non-air-cushion-assisted vessels is the reduction in wetted surface that is achieved through reduced dependence of the vertical force on the action of buoyancy. Following this reasoning, the air-cushion vehicle (ACV) is a strong candidate for high-speed operation, although vessels that are fully supported by an air cushion have limited performance in medium to high sea states because it is difficult to maintain the air cushion. Thus, the SES shows a compromise between the displacement-type vessel and the ACV to deliver a ship that has reduced wetted surface and lower frictional drag but with desirable seakeeping properties.*

The most well-known type of active aerostatic vehicle is a hovercraft. Typically hovercraft are roughly rectangular in planform shape, and fitted with fabric skirts around their perimeter. The skirt serves to retain the air bubble but still permit the vehicle to traverse obstacles, by deflecting the skirt rather than impacting the hard structure.

Hovercraft possess the unique capability of amphibious operation, which is very useful in military application, and may be useful in some commercial services such as ferry service.

Defining Feature: The air bubble.

Sustention: Active Hydro Static (During over-water operation an ACV does in fact displace its weight of water, in the form of an air bubble depressed into the sea surface. It is NOT a -Dynamic sustention vehicle.)

History: Invented by Sir Christopher Cockerell in approximately 1953. Wikipedia has a good article on Sir Christopher.

Seakindliness: Hovercraft only have a modest response to the sea surface up until either the wet-deck slams, or a wave trough causes the cushion to vent. In either of these situations the craft experiences an unpleasant impulsive event. Other seakindliness issues include the so-called "cobblestone" vibration that is induced by pressure pulses coming from the lift fans.

Speed/Power: Because hovercraft have zero wetted surface, they have the lowest drag of any of the AMVs. However, in order to maintain their liberty from the sea they are usually propelled by air screws, which are very low efficiency compared to marine propulsors, especially at low speeds. This mitigates some of the gains in resistance and makes the hovercraft rare for service below about 50 knots.

Air propulsors become more efficient at high speed, and some Military hovercraft do exceed 80 knots. Comfort & Space: The hovercraft's nearly-rectangular planform can make it easy to arrange. The comfort factor is however often reduced by noise and vibration associated with the air propulsion.

Load Carrying Ability: The ACV's load-carrying ability is limited by the maximum air cushion pressure that can be sustained by the skirts. This pressure is exactly equivalent to the draft of a rectangular barge of conventional sustention. Current fan and skirt technology limits this pressure to one to two meters of water equivalent.

Economics (Acquisition & Operation): Hovercraft can be economically built, although they tend to employ lightweight (and thus expensive) structural techniques. Their major cost impact is due to the lift machinery and its associated control systems. In addition, the fabric skirts do wear (something like one millimeter per hour) which necessitates periodic inspection, refurbishment, and replacement.

Alternate Configurations: While most of this discussion has been regarding fast hovercraft, sometimes there are important reasons to employ the hovercraft in low-speed service. An example is the use of hoverbarges in ice-laden or otherwise difficult-to-navigate areas. In such cases the barges are often either towed by winches mounted on land, or even towed by helicopters.

Nomenclature and Terminology: Figure 5.19 taken from the english Wikipedia at: *http://en.wikipedia.org/wiki/Hovercraft* illustrates the relationship of some of the most important components of an ACV, to wit:

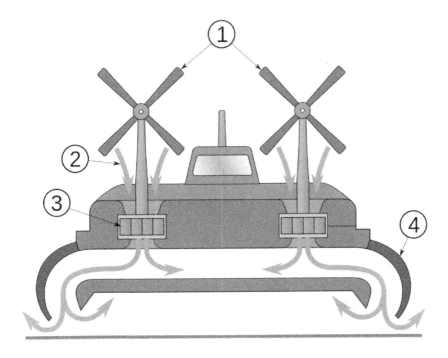

Figure 5.19: A simple schematic section illustrating the defining parts of a hovercraft.

- Propellers

- Air

- Fan

- Flexible skirt

Scalability: Probably unlimited (cube/cube) The two main strengths of the hovercraft are their amphibious capability, and the fact that the absence of frictional resistance may yield very good speed/power characteristics. The key weaknesses are that aerodynamic propulsion is inefficient and noisy, the craft may experience cobblestones, the skirts wear and generate spray, and the craft is difficult to control (having no resistance to sway.)

A few hovercraft pictures follow. An outstanding collection of such pictures may be found at:

http://www.arsp.sojo-u.ac.jp/acv/acv/worldacv/eworldacv.html

Figure 5.20: Sir Christopher Cockerel

Sidewall Hovercraft / Surface Effect Ship / SES

The fully-skirted ACV or Hovercraft suffers from a few impediments, such as the air loss all the way around the perimeter of the craft which drives up the lift power needed. It is also hard to steer, since it has no "grip" on the water, wanting instead to skid sideways like a hockey puck. Further, the use of air screws for propulsion has a huge decrement in net thrust per unit

Figure 5.21: One of the first hovercraft, the Saunders-Roe N-1 (SR.N-1) Note the absence of fabric skirts as are used today.

power, as compared with using marine propulsion, such as marine screws or waterjets.

To overcome these and similar defects, Mr. Alan Ford invented in 1965 what is now known as the SES or Surface Effect Ship, then calling it a "Captured Air Bubble" or CAB craft. The British term for an SES is "Sidewall Hovercraft" and to me this term nicely captures the defining feature of an SES: It has rigid sidewalls, and not skirts-all-'round like an ACV.

The SES is a catamaran-like structure with an air bubble between the hulls. Fabric skirts bridge the gap between the hulls forward and aft, retaining the air bubble. The hulls may be fitted with marine propulsion units. The hulls also provide some roll and pitch restoring force from buoyancy. Sustention: I have listed the SES in the domain of "Active Hydrostatics" just like an ACV. In reality they are actually hybrid craft, wherein 80% (or so) of the lift comes from active hydrostatics (the air bubble) while the remaining 20% comes from the displacement of the sidehulls (passive hydrostatics.)

Defining Feature: A combination of Catamaran and ACV technologies, intending to reduce air leakage, reduce skirt wear and complexity, permit hydrodynamic propulsion, and add hydrostatic stability. Of necessity, an SES is not amphibious like an ACV.

History: As mentioned, the SES was invented in 1965 by Alan Ford of the David Taylor Model Basin (US Navy.) The great push in SES technology development came in the 1970s when the US Navy embarked on

Figure 5.22: The SR.N-1 in overwater operation. Note the large amount of spray created.

an ambitious program to transform the fleet into a "100 knot Navy" by relying extensively on SES ships. The lead ship of this effort was to be the 3000-ton destroyer known then as the "3-K SES". The 3KSES program expended about $500 Million (then year) on research and technology, before finally being cancelled just after the keel-laying of the first ship, in 1979.

Many excellent technical studies and reports were produced during the 3K heyday, far too many to attempt to list here. A good overview of the SES, from those researches, was the paper by Kobitz & Eggington, "The Domain of the SES", Reference [14].

En route to the 3K, the SES program built a series of small test craft designated XR-1 through XR-5, and then two large (80 foot) 100-ton test craft called the SES 100A & SES 100B.

The SES was not adopted for military use, due to considerations of the utility of speed and the evolution of the naval mission, but there have been various resurgences of interest in SES in the decades since the demise of the 3K program.

Seakindliness: Being an 80/20 mix of hovercraft and catamaran, the SES

Figure 5.23: The Saunders-Roe N-4 (SR.N-4) commercial ferry. Note the greatly reduced spray compared to the SR.N-1, due largely to the use of fabric skirts of a design which is still current.

may be considered to be an 80/20 mix of their performance attributes as well. The catamaran hulls respond to waves as do any displacement hulls. The air cushion responds as discussed above. The result is an acceptable ride, that may be better than that of a catamaran providing that the cobblestone effect has been dealt with.

Speed/Power: The SES has somewhat higher drag than a fully-skirted ACV, but this is greatly offset by the reduced lift power requirement and the ability to use more-efficient marine propulsion devices.

Comfort & Space: Generally catamaran-like.

Load Carrying Ability: A little better than catamaran-like, because "there is something holding up the middle of the ship." The limit is that this "something" (the air cushion) has a practical upper limit of about 1-2 meters of draft, and this may be less than the sustention force that one might expect from, say, a barge of these dimensions. Thus the SES does not have the load carrying ability of a barge of similar dimensions, but it is probably superior to a catamaran of similar dimension.

Economics (Acquisition & Operation): The economics of the SES are burdened by the complexity of the lift system. It is difficult to design an SES with less than six engines, for example. (Two propulsion engines, two

Figure 5.24: A Russian AIST class amphibious military hovercraft, generally equivalent to the USN LCAC

Figure 5.25: A Russian LEBED Class ACV

Figure 5.26: The largest hovercraft in the world, the Russian POMORNIK Class at 555 tonnes

lift engines (for redundancy) and two generator engines (for redundancy.)) The skirt systems also add cost, for both acquisition and maintenance.

Alternative Configurations: Nearly all SES are of catamaran configuration with straight-across bow and stern skirts. There were experiments in early days with what were called "partial length sidehulls" wherein the sidehulls were only 50-75% of the length of the raft, and a semi-circular bow skirt was fitted looking rather like the front half of an ACV.

There is also a variant called the SECAT for "SES Catamaran" which was two slender SES side-by-side in a catamaran configuration. Each of the two SES had a very slender cushion, and the SECAT consisted of four sidehulls total, with two cushions.

A variant on the SECAT has been proposed by several designers, which attempts to replace the fabric skirts with rigid structures at bow and stern to contain the air bubble. The nearest to success in this vein that I have seen are the air-lubricated craft developed in Russia. (The interested reader is invited to Google "air lubricated ship" to pursue this subject further.)

Nomenclature and Terminology:

- Sidehull

- Cushion

Figure 5.27: A commercial hovercraft, exploiting the hovercraft's amphibious capability in order to operate in ice.

- Haunch

- Wet Deck

- Skirts

Scalability: No obvious limit (cube / cube)

Strengths & weakness of the SES: Excellent Speed / power characteristics, mitigated by concerns over seal wear, possible cobblestoning, and mechanical complexity.

A variety of photos of SES are given in Figure 5.31 through Figure 5.35. Many of these are taken from the unofficial SES Museum: $http : //www.islandengineering.com/ses_museum.htm$

5.3.6 Active Hydro Dynamic AMVs

None Known.

What might such a craft be? This would be some scheme whereby the craft floats by hydrodynamics, not hydrostatics, but the dynamic effect is produced "actively." The nearest that I can imagine that would satisfy

Figure 5.28: The USN LCAC hovercraft

this would be a "hydrocopter" in which a wing-like rotor keeps the craft up. This would be sort of a hydrofoil, in which the foils are kept moving so that the ship "flies" even when at rest.

A variant would use skis instead of foils, looking perhaps like some sort of fantastic egg-beater. Note, during this excursion into fantasy, how the taxonomy of the Sustention space is helping us to organize our thoughts and indeed helping us to imagine new vehicle types, such as this hydrocopter.

When I wrote the above I was unaware of the existence of any such craft. But thanks to the humor magazine "Deadweight" from the University of Michigan I see that such a craft was not only invented but built. See the following two figures harvested off the internet (Google "hydrocopter.")

5.3.7 Active Aero-Static AMVs

None Known.

Another corner in which I know of no such vehicle. This would be an aerostatic vehicle (e.g. a blimp) but instead of relying on a lighter-than-air gas, it might use a vacuum pump to evacuate its "hull" so that it is buoyed by its displacement in air. Without its fan or vacuum pump (the active component) it ceases to fly.

Early in the design of flying machines some inventors did imagine a vehicle that was buoyed by bronze globes from which the air had been evacuated

Figure 5.29: This picture of an LCAC clearly shows the role a hovercraft can have in shallow-water operation

by pumps, resulting in a displacement of air and thus lighter-than-air flight. Of course, the reality is that the metal globes can not be made light enough to fly in this manner. Will modern materials make such a thing possible in this century? This speculation lies in the domain of Science Fiction and outside this already-far-reaching course.

Intriguingly, it has been suggested to me that a modern hot-air balloon actually belongs in this class, and not in the passive aerostatic class as I have placed it earlier, because the modern balloon requires an active flame to maintain the heat in the lifting gas. Like an SES or Hovercraft if the fans or flame were turned off the vessel will sink down into the liquid (air or water).

Fortunately I do not have to resolve this argument, because however they may be classed these hot air balloons are outside the scope of this book.

Figure 5.30: This picture shows the ultimate in shallow-water: An LCAC on the beach, with the air cushion turned off. Note the deflated skirt visible around the perimeter of the craft.

Figure 5.31: The two 100-ton testcraft SES 100A and SES 100B

Figure 5.32: The SES 100A, the waterjet driven testcraft

Figure 5.33: The SES 100B, the propeller-driven testcraft

5.3.8 Active Aero-Dynamic AMVs

Following on from the Active Hydro-Dynamic AMV, I think that this corner of the sustention cube is occupied by the Helicopter. As such, I am comfortable stating that it is an air vehicle and not an AMV, and thus

Figure 5.34: A commercial SES ferry from Norway

Figure 5.35: The Norwegian Navy SES Patrol Boat SKJOLD

outside the domain of this course.

A helicopter obviously generates its lift through aerodynamics, but this lift is the result of a moving part of the vehicle, not the movement of the whole vehicle. Towing a helicopter forward through the air is not a way to make it fly.[2] Indeed, it is interesting to note that in all four cases the active

---

[2]Interestingly, there is a passive aerdynamic vehicle that looks a lot like a helicopter,

Figure 5.36: The Boeing hydrocopter at rest

Figure 5.37: The Boeing hydrocopter under way

---

called an Auto-Gyro. But this vehicle does not have the ability to hover, and does not actively spin its rotor. Instead, the rotor spins as a result of the craft's forward motion, and acts as an interesting type of passive wing. Google for "autogyro" to learn more.

vehicles are able to hover, whereas the only passive vehicles that can hover are the -static ones.

5.4 Summary of the AMVs in a single table

At the risk of oversimplification, I offer the following vest-pocket summary of the principal AMV types. Note therein that by the term "defining feature" I mean something you can see when you look at the boat, whereas a "performance feature" is some operational capability that you get from the boat - which may be a result of the "defining feature.":

Name	Primary Sustention	Defining Feature	Performance Feature
Hydrofoil	PHD	Foils	Speed & Seakindliness
Multihull	PHS	Multiple Hulls	[-]
Catamaran	PHS	Slenderness	Speed & Volume
Trimaran	PHS	Slenderness	Speed
SWATH	PHS	Submerged Buoyancy	Seakindliness
Hovercraft	AHS	Air Cushion	Speed & Amphibious
SES	AHS	Air Cushion Catamaran	Speed
WIG	PAD	Wing in ground effect	Speed & Seakindliness

6 Hybrids and Weinblums

We have concluded a whirlwind tour of "All The World's AMVs.' Our focus has been upon relatively "pure" or simple versions of the described craft. Now let us consider two cases that may be considered to be orthogonal to the previous classes: Hybrids and Weinblums

6.1 Hybrid Vehicles

Many people have suggested that a benefit is gained by hybridizing, say, half hydrofoil / half SWATH, or a combination between SES and Trimaran, or other similar combinations. Every so often somebody suggests a hybrid:

- ACV/Cat

- Foil/Cat

- SES/Foil

- Planing Hydrofoil

Sometimes it works - but rarely. The critical question to ask when considering a hybrid is:

- Is it solving some particular problem?

- Inadequate stability

- Inability to build a control system

- Inefficient propulsion

- Can you solve it more fundamentally?

It is my contention that in the vast majority of cases hybrids represent not the BEST of both worlds but the WORST of both worlds. Let's consider the question of speed/power performance: In brief, if the lift/drag ratio of concept "A" is 10:1, and for concept "B" is 20:1, then why would I marry A and B? Should I not put all my eggs in the best basket?

Let us explore this further:

"Hybrid-lift" vehicle concepts are those in which two or more primary lift elements (dynamic, static, or powered) are combined, with each element carrying a major fraction of the total lift, not merely trim, stabilizing, or control forces, in a 50/50 (or perhaps 60/40) balance. In connection with a number of recent vehicle concepts, it has been conjectured that hybrid-lift vehicles derive economic or performance benefits from the concurrent use of different types of primary lift, in effect combining the advantages of each. Unfortunately, except for certain specialized missions, it is far easier to defend the contrary assertion: hybrid-lift vehicles are inherently non-optimal for line-haul vehicles, and tend to combine the disadvantages of all lift sources.

6.1.1 Missions And Speeds

Much of the recent interest in high-speed marine vehicles has been motivated by potential applications in line-haul transportation, that is, carrying passengers or cargo over a more or less fixed stage length, at a more-or-less fixed speed. At the end of the spectrum typified by relatively short stage lengths, it is by no means unusual for passenger and even passenger/automobile/truck ferries to operate in the 45-50 knot regime. For transoceanic stage lengths, commercial container carriers and military sealift ships operating in this speed regime are now contemplated, with the expectation of economic viability - or at least military utility - in spite of high unit fuel costs compared with conventional ships 20 knots slower.

This has not always been the case. Not so long ago, very high speeds were considered the province only of combatants - destroyers and patrol craft of various types. Reasons for the change may be found in various areas: economic, geopolitical, and technological. At the risk (nay, the certainty) of oversimplification, it seems possible that future commercial or strategic sealift "ships" with useful payloads in the thousands of tons, will be designed to transit at unprecedented sea speeds, say, in the 50 knot regime or even higher; while future surface combatants may be designed as much for sensitive characteristics (such as low signatures) at "tactical" speeds significantly lower than that of present destroyers. The nature of missions in general, and the role of speed in particular, has changed dramatically even within the last ten years. It is still changing.

Nonetheless, it is important to keep one thing in mind. Many military missions (especially combat missions) involve deliberate and sustained operation in more than one speed regime. Even in civilian life, oceanographic research often imposes two or more speed regimes of importance, as does commercial fishing. By contrast, however, line-haul transit, whether for profit or for sealift, is supposed to be conducted at (or as close as possible to) one economical speed. This speed may or may not always be the original design speed of the ship, as the fuel price dislocations of the past have shown us well enough, but the point is that line-haul is basically a one-speed mission, barring special geographic constraints, such as wash restrictions, or environmental *force majeure*.

Two-speed missions may be viewed as one of the facts of life that drive designers of advanced marine vehicles, in their despair, to consider hybrid sources of lift. For example, an ASW or in-stride mine warfare mission might require sprint (foil-borne, cushion-borne, or on-plane, as the case might be), and search (hull-borne, off-cushion, or off-plane). In some cases the practical difficulties of applying hybrid lift are so severe that a two-vehicle system emerges as a better choice for a two-speed mission.

By contrast, if a mission is truly a one-speed mission, which is what line-haul transit should be, then arguments for and against hybrid lift vehicles should be simpler. But they aren't.

6.1.2 Speed And Lift

In the following discussion, the word "lift" is used not in the aerodynamic sense, but the economic one: "lift" is the force that opposes weight. For vehicles as a general concept, lift may be generated in various ways. For the vehicles of concern here, however, lift is generated entirely by pressures in a fluid, or possibly, two fluids at the same time. Land vehicles (freight trains, for example) are excluded from this class.

It is a custom among high-performance vehicle aficionados to plot measures of vehicle performance versus speed, often with a family of contours for various payloads and/or stage lengths, for a wide variety of vehicle types, in the manner of Von Karman and Gabrielli (see Chapter 7.) The ordinate may be an engineering quantity such as power to weight ratio, drag/lift ratio, some variant of transportation efficiency (for example, hp.hr/ton.mile); or alternatively it may be an explicitly economic quantity such as operating cost per ton mile, required freight rate (RFR), or economic cost of transport, (basically RFR plus a time-value cost on the cargo while in transit.)

As we have discussed, vehicles may be classified meaningfully by which types of lift or sustention are involved (for example, static, dynamic, and powered), and which fluid (water, air, or both) provides how much of the lift. Typically, the classification of types of lift and the fluids supporting the loads are taken at cruise speed. This is an important distinction, because for "takeoffs" and "landings," if any, a different mix of types of lift and fluids (even rubber and concrete) may be involved. For many types of advanced marine vehicles, processes analogous to takeoff and landing are obvious. A representative outline of vehicle types was presented in the foregoing sections.

Using the terminology of the Sustention Triangle, Static lift, we may all agree, comes from differences in static pressure of fluid, acting at different points on a body's surface. Spatial variations in static pressure are the result solely of the weight of a column of fluid. Therefore, although it's a little odd to put it this way, given the fluid density, static lift (buoyancy) comes from gravity. Dynamic lift, on the other hand, does not.

By dynamic lift, generally, we refer to lift produced by the pressure field created by a body's motion through a fluid. It is a semantic difficulty whether the "body" in question moves along with the vehicle, as is the case of the foil of a fixed-wing aircraft or hydrofoil, or along some other path different from that of the vehicle's center of gravity, say, such as the rotor of a helicopter. This difficulty has been solved, semantically, by restricting the term "dynamic lift" to mean lift from a surface moving along with the vehicle, in the sense of a fixed-wing aircraft, and coining the term "powered lift" or active lift to cover other cases, i.e. lift caused by the motion of other parts, rather than the whole vehicle.

"Powered lift" contains its own mysteries, however. It has been argued that several different forms of "powered lift" may be distinguished. To name a few:

- (1) The use of engine-driven moving parts to generate dynamic lift by virtue of their velocity, e.g., helicopter rotor blades.

- (2) The use of mechanical or chemical processes to generate what is basically a static pressure field, e.g., a fan increasing the pressure in an air plenum.

- (3) The use of a jet (even a rocket) engine to develop thrust which supports the vehicle's weight, e.g., an AV-8 at hover.

Now it may be asked why any of these forms of "powered" lift should be regarded as more "powered" than the "dynamic lift" of the wing of an aircraft being driven through a fluid by an engine, and whether each form perhaps deserves a distinct name to provide a convenient reference to its particular characteristics and behavior. For example, one might use the terms "dynamic powered," or "pseudo-static powered," or "vertical thrust," to refer, broadly, to rotors, cushions, or fluid jets, respectively, when used as lift producers. Even then, there may be subtleties that defy concise definitions. For example, what can be said of the translational lift of a helicopter rotor system?

But regardless of terminology, there is little doubt that static, dynamic, and "powered" lift vehicles must operate very differently. Stated glibly, a vehicle supported by dynamic lift will experience stall or an induced drag "crisis" as it slows down from cruise. A vehicle supported by static lift doesn't. However, purely static lift is generally associated with more or less irreducible wetted surface, leading to high drag at high speeds.

Because, typically, all commercial voyages begin and end with a vehicle essentially at rest, dynamic lift must be supplemented, and ultimately supplanted, at some sufficiently low speed by some other form of lift. Stall or its equivalent may not be sudden or catastrophic, but the loss of dynamic lift must ultimately occur, and we better be ready for it. This sad fact can be viewed, in a sense, as the need for landing gear.

Leaving powered lift aside for the moment, and assuming that a vehicle is flying or floating at a constant altitude, total lift can be written in a slightly offbeat form as:

$$L = \rho g A h + \frac{1}{2}\rho A C_L V^2 \tag{6.1}$$

$$L = \rho A[gh + \frac{1}{2}C_L V^2] \tag{6.2}$$

Where:

ρ is the fluid density (for simplicity we assume incompressibility, and that the vehicle is small enough to justify rho as a constant)

A is a fixed "reference planform area" of the body

h is a "reference height" of the body (which may vary with speed)

Thus the first term represents lift due to displacement. The second term represents dynamic lift. C_L is a familiar coefficient which will remain nameless here, in order to avoid confusion, but which is related to the geometry and attitude of the body, and V is the velocity.

Obviously, if two different fluids are involved in lift production, Eq (6.1) should be given an added pair of terms, for example:

$L = \rho_1 A_1 g h_1 + \frac{1}{2}C_L V^2 + \rho_2 A_2 g h_2 + \frac{1}{2}C_L 2V_2^2$

Formally, this equation is complete enough to cover such rare birds as an airship, nose up for aerodynamic assistance, but with its gondola in the water attempting to plane. Interesting as this concept may be, we will restrict the following development to a single fluid, for simplicity. The mass of the vehicle (including everything inside it, even if it's only air or a lighter-than-air gas) is M. Then by virtue of the assumption of level flight:

$h + \frac{C_L V^2}{2g\rho} = M$

In effect, this equation can be a guide to the required area density of a vehicle. To give a perspective in practical terms, A may be considered as the area of a slip, or of a hangar. The "draft" h must be sufficiently small compared to the available water depth or the vertical clear height in an airship hangar. More to the point, conceptually, h, being also related to a physical dimension of the vehicle, has a direct effect on wetted surface, while the second term in brackets does not. One of the reasons why vehicles operating at the water-air interface are economically interesting is that they provide an opportunity to exchange h (and wetted surface) for $C_L V^2$ as the speed changes, with beneficial effects on drag. This opportunity does not exist in the same way for submarines or blimps.

To oversimplify only a little, on a von Karman-Gabrielli plot (ref), the high speed end is the province of successful dynamic lift vehicles, and the low speed end is the province of successful static lift vehicles. Naively, then, shouldn't the middle of the plot be full of numerous successful species of vehicles that derive their lift, at cruise speed, from both sources at once? And if not, why not?

6.1.3 Drag

The foregoing discussion dealt with lift. What about drag? If the product of lift and speed is associated with paying cargo, or at least value added, then the product of drag and speed is associated with fuel expenditure, that is, cash-flow out. If engines and fuel were the only things we had to pay for, then the goal would obviously be to minimize drag for a given lift and speed. Obviously, economics are not quite that simple: we do have to pay for a hull, or wings, as well, and a few other details, but let's accept the simplification for a moment in the interests of the argument.

The challenge is that drag varies with speed, and with the type of lift. At low speed the lowest-drag form of lift is inevitably buoyancy. At higher speeds, for reasons noted above, the situation changes. But what does this imply about hybrids? For any given cruise speed, in principle, there are only two possible situations:

- For practical vehicle configurations, one form of lift will have a significantly better lift to drag ratio than the other

- The lift/drag ratios will be about the same

In situation (1), obviously, we should rely on the form of lift with the best L/D to hold up the entire vehicle weight because that will result in the lowest total drag. In situation (2), which tends to be the case in the speed range for which hybrids are a temptation, we might still want to choose one form of lift, for reasons that are described below.

As just one example, consider a high-speed surface ship: a 3,000 nautical mile stage length with a small payload. Consider a high speed displacement monohull (basically similar to a World War II destroyer in geometry), or a large hydrofoil, each with a first-cut estimated weight of about 7000 tons. The L/D ratios turn out to differ only slightly, and the relative advantage of the two forms of lift depend on the selected speed. The destroyer form is a clear winner at 35 knots and the hydrofoil at 50 knots. The displacement hull form has a volumetric coefficient (displaced volume divided by length cubed) of about 1.6 x 10-3 and a waterline length of 540 feet. Such a hull could reach 50 knots on approximately 278,000 shp. Estimating typical weights of hull, machinery, and fuel (calculating fuel consumption at half load) allows for 588 tons of payload. (The weights are proportioned from recent destroyer data for structures, and assume a constant weight per SHP for machinery similar to that of current US Navy destroyer gas turbine plants.)

Now, we design a 50-50 hybrid for the same speed. Because the hull is supporting only 3500 tons at cruise (with the rest of the weight on the foils), a 540 foot hull is now too long, with excessive wetted surface. Consequently, we reduce the length to 450 feet, with the structural weight reduced accordingly. However, we now have foils, and their associated induced, interference, and parasitic drag. Hydrofoil drag is calculated using lifting line theory, with frictional and pressure drag from Hoerner. Assuming a conservative lift coefficient of 0.3 to allow for takeoff, and using two submerged foils, each carrying half of 3500 tons, and adding their drag to that of a 3500 ton hull, we find that the hybrid's power is reduced to 242,000 SHP. The hybrid is also lighter at 6325 LT. One might suppose that the hybrid may have some advantage over a classical "pure" hydrofoil because there are no struts - it is assumed that the hull would be a low block-coefficient form with the foils attached near the keel.

With payload held constant, the actual power for this hybrid could be reduced somewhat, and the displacement correspondingly reduced, in principle, for the reduction in fuel and engine weights. However, even for the 50-50 hybrid the foils are enormous, the wingspan of a Boeing 727.

Transferring the entire load to the foils, even including estimated strut drags, results in a further decrease to 191,000 SHP. The foils, of course, become nearly twice as big! If we look at the curve of required power against percent of weight dynamically supported, Figure 6.1, we see that it is not a straight trend between 0 and 100%, but that the curve is "convex up." This represents an inherent penalty for having both forms of lift, including interference drag.

Total weight has a similar behavior when plotted against percent of weight dynamically supported, Figure 6.2. The hull required to support

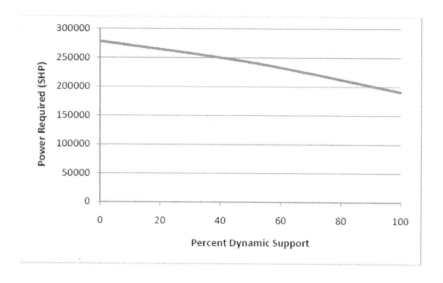

Figure 6.1: Power versus dynamic lift fraction for the example given in text

the vehicle at rest and to contain the payload and engines is considerably smaller and lighter than the 7000 ton destroyer or even the 3500 ton hybrid. (The buoyancy of the foils and struts is considerable and was included in the analysis.)

Similar calculations performed for a speed of 35 knots resulted in another pair of curves, but favoring the monohull. For intermediate speeds, the shape of the curve remains convex upward. The hybrid is always non-optimal at cruise, because as the cruise speed is varied, one form of lift has the better marginal performance, and then the other one does. And because of the upward convexity, hybridism is penalized even when the pure forms are equal in performance. If cruise were the only condition, then we would use one type of lift, appropriate for the speed, and hybrid lift vehicles wouldn't even be a temptation.

6.1.4 Drag Crises

What we have said above is that, for a line-haul single-speed mission, the hybrid form always has worse speed/power performance than the pure form.

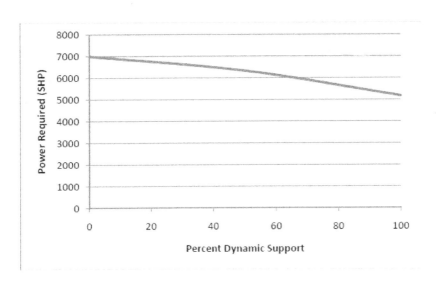

Figure 6.2: Vehicle weight versus dynamic lift fraction for the example given in the text

However, even for a line-haul mission, we must get the vehicle up to its cruise speed, and back down again. In many cases what tempts designers toward hybrids is the need to deal with a drag crisis. The simplest example of this is the planing boat resistance hump: very high drag is experienced at a critical Froude number. A similar drag crisis is experienced by dynamically supported vehicles flying at low speeds, just above stall. This may become the point which determines the vehicle's installed power. The engine power required to get over hump may dictate the top speed, rather than the other way around. But even in this case, the passage through the drag crisis is accomplished within the same mode of sustention, not be recourse to a hybrid mode.

An example of this principle in success is a commercial jetliner. It does make provision for passing through a low speed regime in flight via flaps and slats, but not this by augmenting its dynamic-lift performance, not by having recourse to a hybrid lift mode. It does have a secondary lift-producing system (wheels and tires), which are used for an even lower speed regime, but not in flight. A typical weight fraction for landing gear on an airliner is about 3 percent.

The evidence suggests that the designer is better off finding some way to power through the drag crisis, than trying to develop an efficient hybrid. Alternatives to hybrid lift do exist for dealing with humps: additional power is one. JATO, catapults, and staged vehicles are other examples of systems used to assist single-lift craft to flying speeds. While alternatives such as these have not been widely used in the marine vehicle setting, the precedent exists: early human-powered hydrofoils were launched by a slingshot. The cruise plant (the human rider) had then merely the task of maintaining flight, as opposed to working his way through takeoff and drag hump. The design question that dominates all such tradeoffs is weight fraction: How much weight (lift capacity) must be dedicated to dealing with drag crises? What is the acceptable weight penalty to pay for a landing gear? A comprehensive answer to this is outside the scope of this chapter, as it will differ according to the mission of the vehicle,and the requirements of supporting infrastructure. There are, as yet, no catapult-equipped quays or paved runways on harbor bottoms for the benefit of hydrofoil takeoffs and landings.

However, some observations may be made which may be seen to be obviously axiomatic: the weight of the "landing gear" is subtracted from the payload carrying ability of the craft. Without wheels the airliner could carry a few more passengers or a bit more fuel, or be equipped with a bit less power.

This, of course, immediately leads into the design spiral: eliminating the landing gear will reduce the weight of the craft which will reduce drag which

Figure 6.3: A bad planing boat but a good hydrofoil?

will permit reductions in power which themselves reduce weight and so forth, until a new design convergence is found at a smaller lighter airplane.

In the maritime example we may consider the "landing gear" of the hydrofoil. This is the ship-shaped main hull, which supports the craft during takeoff and landing. It is interesting to compare the ship-like hulls of some hydrofoils with the unusual hulls of the commercial JetFoils - especially when we consider that the JetFoils were developed by an airplane company. Does this choice of hull shape represent an attempt to make the hull form function more vestigial, more of "merely" a landing gear, where other hydrofoil designers have chosen to make hulls which are good ships in their own right? (Think how good a ship a foil-deprived hydrofoil might be. Compare this to how terrible a motor coach a wingless jetliner would be.)

6.1.5 When Hybrids Work

There are hybrids that work. A few examples are:

- Foil assisted vessels - Eliminate the need for a foil control system

- Tail-draggers

- Foilcats

- SES (Hybrid Cat and ACV) - Reduces air leakage and provides for use of marine propulsors

- Semi-SWATH Catamarans - Reduces ship excitation by waves, without demanding active control

The author's contention is that these are cases where the presence of the "other" lift system brings to the table some capability or solves some problem that is caused by the "good" lift system. For example, a fully submerged hydrofoil is optimal from a lift/drag point of view, but it demands an active ride control system. A tail-dragging configuration (with surface-piercing bow foil) can eliminate this need. An ACV is superior to an SES from consideration only of drag, but by adding about 20% of passive hydrostatic support the ship can greatly reduce air leakage and lift power, and can better accommodate marine propulsors (which are substantially more effective than air propulsors.)

6.1.6 The V-K Gap: Physics Or Just Lack Of Imagination?

When a typical von Karman-Gabrielli plot is made using a performance variable as the ordinate, such as Kennel's transport factor (see Section 10.3), it is difficult to point out the so-called "von Karman gap." The transport factor of the best types in each speed regime seems fairly well behaved, and the curve proceeds relatively smoothly from one type of lift across to the next. When economic performance is plotted, however, the V-K gap tends to show up as a region where the economics of the best examples fail to follow the progression from low speed and low cost per ton-mile to high speed and high cost per ton-mile. They are all worse.

But if the gap is real, where does it come from? I believe that it comes from the nature of lift production. The V-K gap is an unavoidable consequence of one form of lift that experiences stall or an induced drag crisis, and another form that experiences no induced drag crisis but which has a drag penalty at high speed due to excessive wetted surface. There are classes of vehicle that do not seem to have a gap (at least within practical speeds). Freight trains do not. Further, it seems possible that on other planets, with different values of g, or with fluid densities and viscosities widely different from those of water and air, the V-K gap might not be so prominent. But we have to deal with the planet we've got.

I further believe that attempts to discover vehicles which operate in the heart of the V-K gap, and still make money, are long shots. It seems that there is more to be gained by concentrating on placing the vehicle wholly in one regime or in the other, and then minimizing the weight fraction expended on "landing gear."

6.1.7 Conclusion

In summary, the author's beliefs are:

- A hybrid vehicle combines, not the best of both worlds, but typically the worst of both worlds

- Some hybridization is required for any dynamically supported vehicle (landing gear are a necessity)

- The secondary form of lift should be made as vestigial as possible. The best hybrid will be the least balanced, i.e., a 90/10 vehicle is superior to a 50/50 vehicle

- Application of this thought to modern marine craft may lead to radically new types of vehicles (What does a hydrofoil look like when all other modes of support have been minimized?)

6.2 What about Weinblums? Why must ships be symmetric?

In Section 6.1 we considered the question of why ships should rely principally on one form of lift. Now let us tackle another interesting question: Why are ships laterally symmetric? Are there cases where an asymmetrical ship would have some advantage over a symmetrical one? There are very few asymmetrical ship concepts that I am aware of; in fact, I know of only three. One is the polynesian outrigger canoe. Another is an unusual oblique icebreaker concept developed recently by Kvaerner Masa Yards. see US Patent 5996520 A.

More common in the AMV community is the asymmetrical ship called a "Weinblum." The Weinblum name was coined by H. Söding in 1997 [15] as a combined tribute to Dr. Weinblum and a reference to the asymmetry of a grape vine - called a "weinblum" in German.

Herr Söding studied the effect that would result if the two hulls of a catamaran were staggered longitudinally. He found that at some stagger ratios there could be quite dramatic wave cancellation, as illustrated in Figure 6.6.

Naval Architect Paul Kamen et al in [16] wrote: "Another possible benefit of asymmetrical multihulls is manipulation of the wake waves. It may be

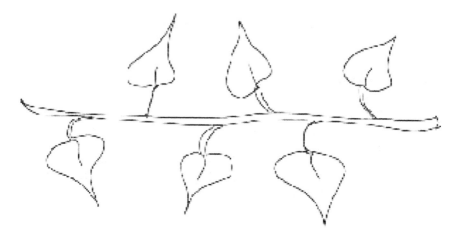

Figure 6.4: A sketch of a grapevine, or "weinblum." Note how the leaves are staggered port-starboard-port-starboard etc.

possible to build a vessel that leaves waves on only one side. This violates the answer to the classic trick question, 'what happens if you tow a half-model down the middle of the tank?' "

Applications for such a configuration are of course limited, but intriguing: Consider a large lake lined with waterfront properties, subject to wake damage. A circular ferry service, always circling the lake in the same direction, could benefit from an asymmetrical multihull that only makes significant wake waves on the offshore side. (There might have to be one boat for the clockwise route and one for counter-clockwise.)

Figure 6.5: Herr Dr. Georg Weinblum

Figure 6.6: A plot of the wave pattern from a Weinblum hull, consisting of two identical hulls staggered longitudinally

7 Performance Metrics

As designers of advanced marine vehicles, we are in the position of explorers. And as explorers we need skills in map-making and path-finding. How shall we determine that we are on the track of a good idea? How do we estimate which direction to take to improve our design? How does our design compare to others? The von Karman, Kennell, and McKesson techniques presented herein represent exactly that type of skill, and we shall spend a few lectures acquiring them.

If I may be permitted a metaphor, I would liken this to learning to drive a car: Designers of conventional ships have the benefit of a well-established network of roads, streets, and highways, and maps and other navigational aids, and in the words of Captain Ron "If we get lost we can just pull in somewheres and ask directions."

As designers of AMVs we are not so fortunate. Our Drivers Training course begins with instruction on how to use a hatchet to clear a path through the brush, how to test a stream to see if it's shallow enough to cross, and maybe even how to find our position using the stars. Imagine if motor vehicle training in America had to begin with those subjects! As AMV designers we are in the shoes of Lewis and Clark. The tools presented in this next unit are the tools for exploring a frontier.

These tools are interesting for two separate purposes: Design and Analysis:

- During *design* these techniques can serve as "small scale charts" of the design space, to tell you where you might profitably look for a solution - and where you should not bother to look.

- During *analysis* I call them "lie detectors." There are plenty of poor solutions floating around in the market, usually just design proposals. However sometimes poor solutions get picked up because of strong marketing efforts. Critical thinking is needed, and tools for critique are essential.

One thing you will note is that these tools are not necessarily specific to marine vehicles. Learn to value the way other people think about similar problems, e.g. aerospace engineers, land-vehicle designers, etc.

And finally, note that these tools are not definitive: AMVs are exciting because there is no single right answer - many good solutions exist. Therefore please argue with me.

7.1 Von Karman / Gabrielli curve

The classic treatment of transport efficiency was a seminal paper by Theodore von Karman in 1950 entitled "What Price Speed?" [1]. The von Karman methodology that I will summarize below presents a map of the cost of speed, providing a technique for understanding what is really involved in making a vehicle faster. I then move from this to discuss the trade-off between the *cost* of speed and the *value* of speed. This conceptual model, based on the principles of economic science, was developed by Mr. Victor Norman in Norway [17].

Dr. Theodore von Karman (Figure 7.1) was, simply, a genius. His contributions to engineering are too numerous to list. In 1950 he published a conceptual relationship for comparing the effectiveness of competing vehicles [1]. His parameter of interest was the specific power required for propulsion of the vehicles at their design speeds. This specific power is a non-dimensional quantity of "power per unit transport." Von Karman's "specific power" is the inverse of the modern "Transport Efficiency."

Efficiency metrics are often designed as a fraction with the "goal" in the numerator and the "cost" in the denominator. The question then is what are the "goal" and "cost" of transport? The answer is that the "goal" is to move some weight at some speed, and the "cost" is the power required to accomplish this. Thus:

Transport Efficiency: $\eta_T = Weight * Speed/Power$

Note that this fraction is non-dimensional, since pwer = force x velocity.

Von Karman collected a database of examples and plotted the best of each class of vehicle. His graph is reproduced in Figure 7.2. He then observed that there appears to be a line, diagonal on his log/log axes, that defines an apparent frontier or limiting value. He also observed that there were no vehicles in the range of 100-200 mph that lay along this line - that there was an apparent "gap" in our ability to accomplish the ideal level of performance. This region is called "the von Karman Gap."

I have repeated the von Karman exercise several times, asking classes of undergraduates to collect data on the speed, weight, and power of a variety of vehicles. I have collected all of this data, without much "scrubbing", and plotted it in von Karman's style - see Figure 7.3.

Compare the "cloud" of data in Figure 7.3 with von Karman's original curves. We see remarkably similar characteristics: There is a "lobe" of spots where von Karman has noted "merchant ships." There is a lobe of data at high speeds near the aircraft. There is an arguable "frontier" that could be drawn as a diagonal on the lower right. And there is a gap around 100 knots.

Figure 7.1: Theodore von Karman

7.2 The Value of Speed

The von Karman diagram shown above presents a map of the cost of speed, providing a technique for understanding what is really involved in making a vehicle faster. Mr. Victor Norman in Reference [17] expands from this beginning to construct a conceptual model of the balance between cost and value, as applied to high speed shipping.

The economics of fast shipping is the balance between the cost of shipment and the value of shipment time. In von Karman's language, the cost of speed compared to the value of speed. This doesn't matter whether we're talking about 40 knot shipments or 10 knot shipments, the principle

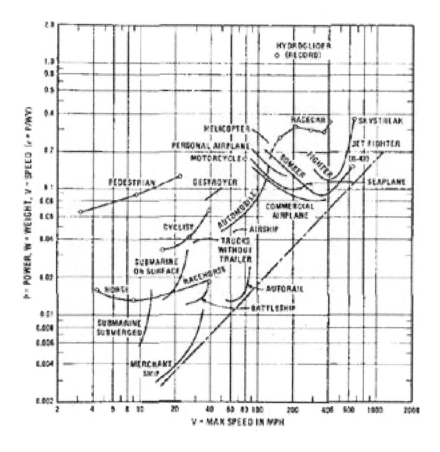

Figure 7.2: Von Karman's 1950 graph of Transport Efficiency [1]

of balance is the same.

7.2.1 The Cost of Speed

Speaking first of the cost of speed, we know that it obeys the laws of physics. We can say a few things in general about the cost of speed. We know that it does not go through zero. There is some cost associated merely with building a metal box around a cargo. We know that it rises with speed. (These things are obvious to us but in fact, frequently we see people who will ignore a few of these truths.) Figure 7.4 presents a conceptual graph of these truths.

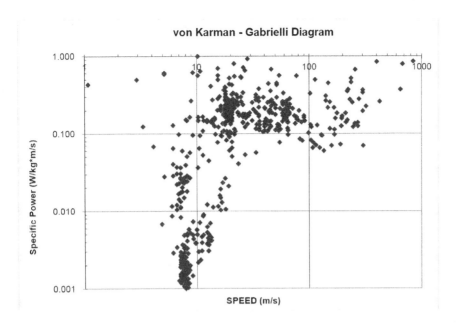

Figure 7.3: Von Karman data collected by a class of undergraduates

7.2.2 The Value of Speed

In the same way as the cost of speed, we know a couple of key points about the value of speed. There is some minimum useful speed below which I need not bother to ship the goods. If you can't move them at least this speed, it's no use to me. And at the other end of the spectrum, there may be some speed above which additional time savings have no merit, perhaps because another link in the transportation chain becomes saturated, perhaps the market cannot absorb the goods that fast - whatever the limit may be. This can be sketched as a graph as in Figure 7.5. And between these two corners there is a region which I have depicted as a straight line.

And, of course, in the perfect world the value of speed and the cost of speed meet at some unique point and this is the point at which I ship the goods. It's usually not quite this clean.

In all of these discussions, I have made the implicit assumption that ship speed is the speed that I'm talking about. This, of course, is because I'm a ship designer. Naturally I think that my part of the transport chain is the most important part. But in fact, ship speed is only one part of total throughput. There is no point in hurrying the ship to arrive at the port just at the time that dock workers quit for the day. There is no point in

121

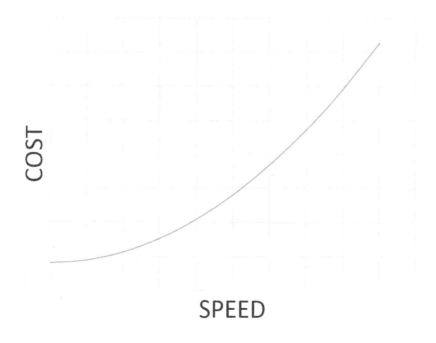

Figure 7.4: The unarguable truth of the Cost of Speed

hurrying 10,000 TEU to a port that can only off-load them slowly, limited, let us say, by land space available. In my portion of the United States, one of the attractive types of speed is frequency of service. If your cargo arrives at certain terminals in Seattle, we can put it on a barge leaving for Alaska today. We have another barge leaving for Alaska tomorrow morning, another one tomorrow afternoon and so on. The frequency of service is such that a 10-knot ship technology sailing daily results in faster service than a once a week service by a 30-knot ship.

7.2.3 Technology Affects Cost

Ship technology affects the cost of speed, that's why I dwelt so long on the sustention diagram. The curve of cost of speed that I showed a few moments ago is, in fact, the bottom of a family of curves - see Figure 7.6. If I really want only to house the cargo and move it very, very slowly, clearly a barge is the right hull form. If I wanted to move at conventional ship speeds, then a conventional ship is the right tool for the job. If l want to move at speeds

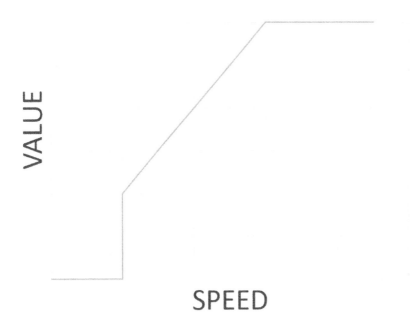

Figure 7.5: The conceptual sketch of the Value of Speed

of 50 or 100 knots, an SES or other form might be the right hull for the job. In fact, the cost-of-speed graph begins to look a lot like von Karman's graph.

7.2.4 Cargo Affects Value

If the sustention technology affects the cost of the speed, then surely the nature of the cargo affects the value of the speed. There are a few classes of cargo that are immediately obvious: Time insensitive goods. The last time a major technology changed the speed of shipping was when steam replaced sail. The last cargo to move in sailing was nitrate: bird guano. It wasn't at all time sensitive. In fact, it didn't much matter how long it took to get it to the market as long as you got it there eventually, and nitrate was the last cargo shipped in sailing ships. On the other hand perishable cargo has an obvious time sensitivity. Kiwi fruit and star fruit were mentioned at one

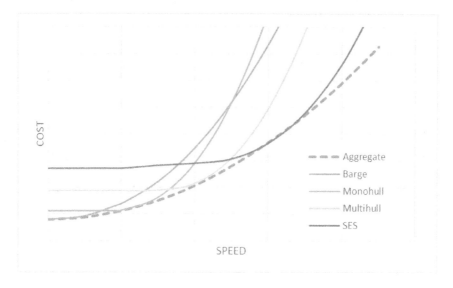

Figure 7.6: The cost of speed depends upon the technology selected. (NB: Lines depicted are notional only.)

conference as examples of highly perishable cargoes. In some of these cases we take other technological tools and we try to change what I will call the "perishableness" of the cargo, but I believe that the hands of nature can only be pushed so far, and that there comes a point at which speed is still the only tool for getting the cargo to market before it spoils.

There are time sensitive cargoes whose sensitivity is purely economic in nature. Computers, if delayed long enough in shipment, might in fact reach obsolescence or at least lose market value substantially *en route*. And if that example is a little too extreme, I know of one case of an American automobile manufacturer who built automobile bodies in Italy and flew them to the United States for final assembly. Not because automobile bodies can't be shipped in containers but because the customers can't wait. They want to order the car now and have it today. We can't quite get to today, but air shipment definitely had economic merit.

7.2.5 Economics Affects Both

If technology affects the cost of speed and the nature of the cargo affects the value of speed, then obviously the world's economy affects them both. The cost of speed was presented as being a curve which obeys the laws of physics. What the curve truly is, is a curve of the energy consumed vs.

speed. In the same way the value curve (or the linear region of it) is a curve of the cost of time. The slope could be expressed in terms of dollars per hour. Dollars per hour is also of course the measure of labor rates, which measure the price of time for people. How much must you pay a man to wait or to do anything he doesn't want to do? In the same way how much must you pay to tie money up? That's interest rates. That is the price of time as it affects cargo. Labor rates are the tool for measuring the value of time to a person, interest rates are a tool for measuring the value of time to a cargo.

Now I need to quickly say that I'm a naval architect and not an economist - I've taken this argument from Reference [17]. But I think it's very important to understand the economic principles so that we can apply them to our own situation. What I hope that you leave with, is not a forecast of tomorrow, but the ability to make your own forecast based on the economic realities of your own state.

Figure 7.7: About fifty years of "value of time" data for people, corrected for inflation, from the Wikimedia Commons.

Labor rates are the measure of the cost of time for a man. Figure 7.7

shows the labor rate of the 20th century, corrected for inflation. It's not visible in this graph, but other data I have seen suggests that the peak in the 1960s was the culmination of an eight-fold rise in the real value of labor.

At the same time oil prices may be taken as the cost of energy - in other words, the cost of speed. Over the majority of the 20th century, we saw a net decrease in the price of energy, until the 1970s - see Figure 7.8.

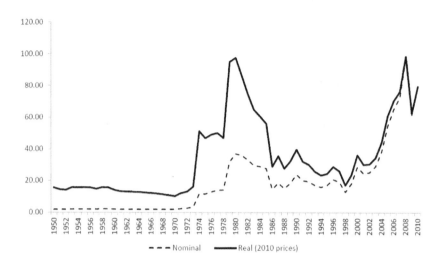

Figure 7.8: Nearly a century of "Cost of Energy" data, to compare with the previous graph

The result was that up until the 1970s, we saw a 16-fold increase in the importance of time compared to the price of energy. This is seen in the ratio of those two curves - labor rate divided by oil price. When oil prices are low compared to labor rates, speed is valuable, speed is marketable. Across the first 70 years of the 20th century, speed was important, it was marketable. Across the first 70 years of the 20th century, we saw tremendous improvements in the speed of transporting people.

Remember that this curve applies to people because the figure for the value of time that I have used is labor rates. Think of the technologies that were developed in the 1960's, in the post-war period, in the between war periods, tremendously important technologies in the transport of men. Think of the lack of such developments that occurred during the 80's, during the oil shock periods. There were no supersonic transports conceived and promulgated the way the Concorde was just a decade earlier.

Figure 7.7 was based on the value of time for men. Let's look at the value of time for cargo. The measure again is interest rates, plotted in Figure 7.9. The window here is smaller. This is strictly the post-war window. This is US data, but in fact the world followed the same trend. A rising interest rate period, a lower interest rate in the early 1970s, a sharp rise to the 80's, and I'm sorry this data doesn't go up to today. But I can still compare it to the oil prices for the same period. Again, in the 1960s, the value of time to cargo was high compared to the price of energy.

Figure 7.9: The value of time for goods (interest rates) for 50 years of US history (Source: DollarDaze.org)

In the 1960s, speed paid. Speed was marketable. SL-7s were developed. A lot of high speed freight developments took place. A lot of ports in the world saw their airports become more important than their seaports during this time. During the oil shock period, and into the 80s, an opposite trend existed. While the value of time (interest rates) may have been high, the cost of energy was even higher. There was less development of new speed technologies, but the two curves were still pretty much parallel to each other. What has happened in the 90s, and will happen in the future? It's a good question. Are these curves crossing? Is speed becoming more important than time once again? Some people think so, some people think not.

7.2.6 What Does the Future Hold?

What I want to do is to give you a tool so that you can use data from your own situation to make your own forecasts and understand the pressures affecting your design decisions. I can't, however, resist the temptation to do a little forecasting and a little crystal ball reading. World wage rates are probably stable. Interest rates are low. Oil prices, however, seem to be rising after taking a dip a year or so ago. If the value of time is stable and the cost of speed is rising, then the value of speed is declining. I am a fast ship technologist who believes that cargoes are getting slower.

Why then are people developing fast ships? Because airplane designers suffer from the same economics. I believe that there are cargoes that may move out of air freight and into sea freight if there is a sea freight technology just one step on the ladder below them. In the same way, I will not be surprised to see cargoes move from, say, conventional ships to barge lines, from the same pressures. I understand that in the 1960's a lot of ports saw their airport become more important than their sea port. If the prediction is right, we may see a correcting trend in the coming decades. Reducing the value of speed may drop certain cargoes out of air freight.

Again my analogy is stepping down the ladder. There may be an economic pressure to step to one rung lower. That only works if there is, in fact, a rung one rung lower. That's why so many shippers are developing high speed ships, to put a viable alternative just below the air freight alternative.

I believe this same effect will show up in other cargoes as well. The high speed ship technology that I mentioned applies to these other areas as well. We can imagine SWATH barges making it possible to ship barge cargo into rough seas, seas that can now only be served by conventional ships. We can imagine catamaran barges making possible barge traffic at a speed of say 18 knots. We might even imagine routes that are not being served now seeing traffic because, if the economic pressure for speed declines, a route that is "too slow" becomes "fast enough."

- If the cost of speed is low, and the value of time is high, then speed sells.

- But you've got to know what you're moving - people or merchandise

- Pay attention to global economics - it directly affects your business

- You now have a tool to know which direction the effect will be

- There are other competing technologies than just your classmates'

- Fast ships may be better than slow airplanes

- Are we ship designers, or are we transportation providers?

7.3 Kennell Transport Factor

We have seen how von Karman's specific power methodology can be used to model economic truths. Let us now see what other insights can be gained from this line of reasoning.

In 1998 Dr. Colen Kennell introduced a variation on the von Karman/Gabrielli metric and named it Transport Factor [2].

The motivation for the study of Transport Factor is the same motivation that we have encountered a few times previously, and that we will encounter again: To try to make sense out of too-many options. Dr. Kennell quotes Kenneth S.M. Davidson [18]: *"In these days of rapid change and expanding possibilities, the need for a clear over-all view has already made itself felt and seems likely to grow greater. Not long ago the problem arose of assessing as far as possible in advance the potentialities of a proposed novel type of transport craft that would have characteristics lying in the vast region between merchant ships and commercial airplanes."* To put it more colloquially: "The challenge is in making sense of it all!" But through the lens of Transport Factor (or other similar devices, as we shall see) we learn several important conclusions:

- There is structure to the universe

- High speed ships are different, but fit in with conventional ships

- Yardsticks/metrics are useful for establishing familiarity at the highest level

- Parametric assessments can provide useful insights

Let us now delve into the details of the Transport Factor, and see how these conclusion emerge.

7.3.1 Transport Factor Defined: TF is L/D

Kennell's Transport Factor is defined as:

$$TF = K \frac{W}{SHP/V_K} \tag{7.1}$$

Or:

$$TF = \frac{W(lbs)V(fps)}{550SHP(hp)} \tag{7.2}$$

In the latter case, with English units, the constant "K" becomes $(1/550)$ and has the effect of converting the power (SHP) into a pseudo-resistance. "K" is $(1/326)$ when V is in knots, as in Kennell's original definition, while the W and SHP remain in pounds and horsepower, respectively.

Let's look at this relationship a little further. Consider the case of a ship. Using English units, we recognize the following relationships:

$$EHP = \frac{R_T(lbs)V_K(knots)}{326} \tag{7.3}$$

and

$$SHP = \frac{EHP}{OPC} \tag{7.4}$$

Thus the TF component $\frac{SHP}{V_K}$ is found to be:

$$\frac{SHP}{V_K} = \frac{\frac{EHP}{OPC}}{V_K} = \frac{R_T}{326(OPC)} \tag{7.5}$$

thus:

$$TF = K\frac{W}{\frac{SHP}{V_K}} \tag{7.6}$$

$$TF = \frac{1}{326}W\frac{326(OPC)}{R_T} \tag{7.7}$$

$$TF = OPC\frac{W}{R_T} \tag{7.8}$$

In other words, for an OPC $= 1.0$, the TF is simply the ship Lift/Drag ratio, W/R_T. This has important implications in later discussions.

7.3.2 Transport Factor Decomposed

A key element of the Transport Factor formulation was Dr. Kennell's insight that the TF of the total system can be decomposed by decomposing the weight term in the numerator. Thus, as Kennell proposes:

Since:

$W = W_{ship} + W_{cargo} + W_{fuel}$

Then in the same manner:

$TF = TF_{ship} + TF_{cargo} + TF_{fuel}$

Breaking the system down into these three parts can give useful insights into "where the work is going" in the ship. That is to say, a ship that has a very large fraction of its TF devoted to TF_{fuel} is expending its energy carrying its own fuel around, leaving very little for the carriage of cargo.

7.3.3 Study of Size and Slenderness Effects

Let's begin by getting familiar with Transport Factor in its general behavior. Kennell provides a collection of data very similar to von Karman's, reproduced in Figure 7.10. Dr. Julio Vergara presents a very similar trendline, reproduced in Figure 7.11 (private communication.)

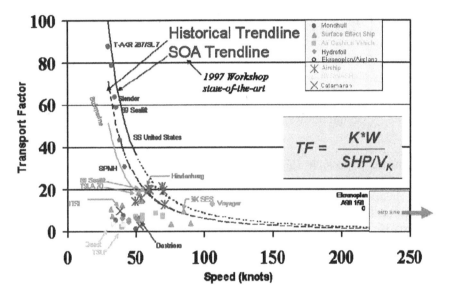

Figure 7.10: Kennell's TF trendline, from [2]

We will present and discuss many such trendlines, but first let us make a cautionary statment about them. It is vital to understand that these are observed trends, not laws of physics. That is to say that we, the authors of these curves, notice that there appears to be a frontier which nobody has yet crossed. This does not mean that this frontier is real, it does not mean that it will never be crossed, indeed it does not even mean that it can necessarily be approached or arrived at. It is merely an observed trend.

The first thing to note in these trendlines is the tendency with speed: It is only possible to attain a high TF at a low speed. If speed increases, then the total TF appears to necessarily fall. But speed isn't the only

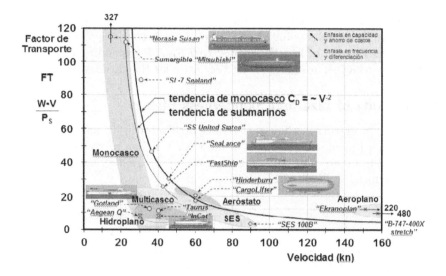

Figure 7.11: TF Trendline proposed by Dr. Julio Vergara (Chile)(Private Communication)

determinant of possible TF - size is apparently important as well. Kennell provides some data on small fast ships (Figure 7.12), and as we see not all of them are able to approach the TF "frontier." So speed alone is not enough to determine the "state of the possible."

Next, Kennell considers the effect of slenderness. In 1998 he conducted a study of 10,000 ton monohulls, of different slenderness ratios [2]. The results are plotted in Figure 7.13. It is clearly evident that increasing the slenderness of the hull helps the hull to approach the TF "state of the art." In fact, as mentioned in the earlier chapters of this book, the desire for slenderness is the very *raison d'être* for several of the advanced marine vehicles, most notably the catamaran and trimaran. The catamaran and trimaran forms are ship types that make high slenderness feasible, by finding another solution to the stability problem.

Kennell's 1999 paper [3] fully recognizes this, and went on to study cat and tri hulls, both having slenderness of about 10, but of different size. They found that the larger the ship, the higher the TF - see Figure 7.14.

7.3.4 Fuel Consumption - TF_{fuel}

Kennell's next step was to explore what we can learn from the weight breakdown of the TF, and in particular what TF might have to say about

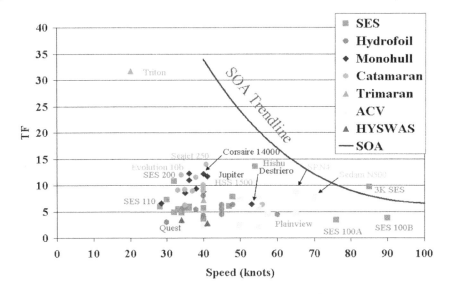

Figure 7.12: Kennell's experience data for small fast ships. Note that not all of them are able to arrive at "State of the Art" performance

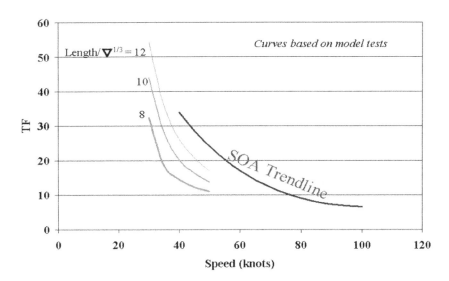

Figure 7.13: Kennell's data on the effect of slenderness, from [3]

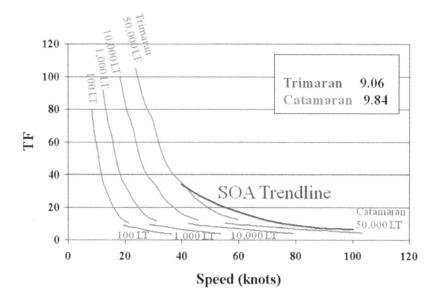

Figure 7.14: Kennell's graph of the effect of size upon attained TF

relative fuel consumption of various craft. Recall that $TF = TF_{ship} + TF_{fuel} + TF_{cargo}$. Kennell reports (Figure 7.15) historical data that shows a linear trend between range and the amount of TF that is spent on fuel, and this relationship does not appear to depend on ship size. (The reader may not know it, but the monohull ships plotted in Kennell's data range across about two orders of magnitude of displacement, from the ARS to the SL-7.)

Why would this be true? Let's look at how fuel loads are calculated for US Navy ships, and all of a sudden this surprising trend will become clear. US Navy fuel loads are calculated by the procedure given in DDS-200-1. In brief, the procedure and assumptions are as follows: Assumptions:

- constant displacement

- high installed power

- endurance speed = service speed ˜full power

- hotel load fuel is negligible (˜5%)

- no burnable fuel at end of voyage

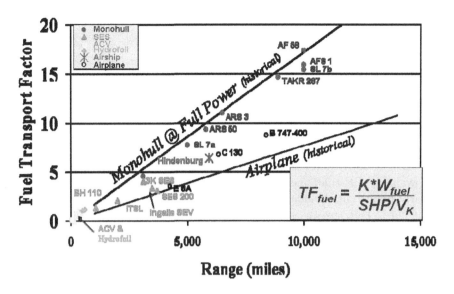

Figure 7.15: Kennell's historical data on TF fuel trends

This gives the relationship:

$$FW = SFC * P_{end} * \frac{Range}{V_{end}} * \frac{(1.0 + k_{fr})(1.0 + k_{pd})}{(1.0 - a_t)} \qquad (7.9)$$

Where: FW = Fuel weight
SFC = Specific Fuel Consumption
P_{end} = Average endurance power
V_{end} = Endurance speed
k_{fr} = Fuel rate correction factor
k_{pd} = Plant deterioration factor
a_t = Tailpipe allowance
Since:

$$TF_{fuel} = K * FW * V_K/SHP \qquad (7.10)$$

Then, if $P_{end} = SHP$ and $V_K = V_{end}$, we can see that a calculation of TF_{fuel} according to DDS 200 will result in the SHP and V_K canceling out of the two formulae, leaving:

$$TF_{fuel} = K * SFC * Range * \frac{(1.0 + k_{fr})(1.0 + k_{pd})}{(1.0 - a_t)} \qquad (7.11)$$

Or approximately:

$$TF_{fuel} = .003622 * SFC * Range \qquad (7.12)$$

135

where:
$$0.003622 = K * \frac{(1.0 + k_{fr})(1.0 + k_{pd})}{(1.0 - a_t)} \tag{7.13}$$

in English units, with SFC in pounds per horsepower-hour and $Range$ in nautical miles, where $K = \frac{1}{326}$, $k_{fr} = 5\%$, $k_{pd} = 10\%$, $a_t = 2\%$.

Which is exactly the linear relationship revealed in the data.

7.3.5 SFC effects

The insight above tells us that not only is TF_{fuel} (i.e. the amount of the available TF that must be spent on carrying fuel around) dependent linearly upon range, but it is dependent linearly upon SFC as well. If we could cut in half, say, the fuel burned per kilowatt of power produced, then we would similarly cut in half the amount of TF that we have to expend on fuel carriage, as opposed to carriage of cargo.

A 1997 sealift technology workshop published a collection of state-of-the-art values for SFC of various machines [19]. Kennell took that data and applied it to the $TF_{fuel} = f(range, SFC)$ formula to produce the trend lines in Figure 7.16.

7.3.6 Fuel Weight Fraction

If the TF of the entire system can be predicted from the State of the Art line, and the TF expended on fuel is a linear function of range and SFC, then how much of the total TF is usually expended as TF_{fuel}? This was Kennell's next inquiry.

His data (Figure 7.17) shows that this is dependent upon two things: Range and speed. High speed vehicles consume more of their TF budget carrying fuel, than do low speed vehicles. Note that this is not the numerical amount of TF, but rather the fraction of total TF that is spent on TF_{fuel}. Since TF_{fuel} does not depend on speed this trend is occurring because the denominator, TF_{total}, decreases with speed.

7.3.7 Emptyship Weight - TF_{SHIP}

Having now established some trends for total TF, and some relationships for the amount of TF that is spent on fuel, how much of the TF is spent just upon the empty ship? Recall once again:

$$TF = TF_{ship} + TF_{fuel} + TF_{cargo} \tag{7.14}$$

$$TF_{ship} = TF - TF_{fuel} - TF_{cargo} \tag{7.15}$$

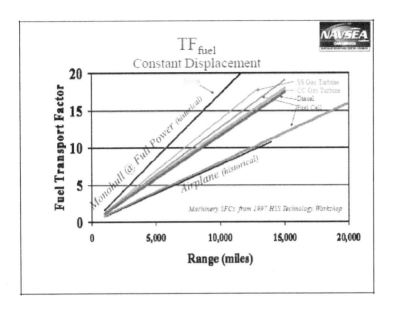

Figure 7.16: Kennell's plot of the relationship between propulsion technology and TF_{fuel}

$$TF_{ship} + TF_{cargo} = TF - TF_{fuel} \qquad (7.16)$$

$$TF - TF_{fuel} = K * (1 + W_{ship}/W_{cargo}) * (W_{cargo}/SHP) * V_K \qquad (7.17)$$

In his 2001 paper "On the Nature of the Transport Factor Component TF_{ship}", [20] Kennell used the above derivation to study the trends for both W_{ship}/W_{cargo} and W_{cargo}/SHP - the two key terms in the final TF(ship+cargo) relationship. The key result is given in Figure 7.18. This shows that there is a direct relationship between these two parameters, where one trades against the other.

The next relationship regarding emptyship weight that Kennell identifies is the relationship between this weight and the density of the deadweight.

Deadweight density is defined as:

$$Deadweight Density = (cargowt + fuelwt)/(cargovol + fuelvol) \qquad (7.18)$$

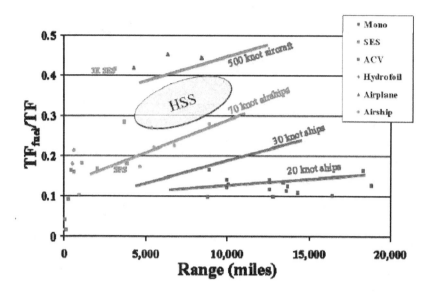

Figure 7.17: Kennell's finding on the proportion of TF devoted to fuel, as a function of speed and range

In Figure 7.19 Kennell has a very clear graphic which shows the trend of Deadweight Density as a function of ship type or ship mission. He doesn't include them, but I suspect that Ore Carriers may have the highest Deadweight Density of any common merchant ship.

What Kennell has found is that Deadweight Density is a useful predictor of the total ship / light ship weight relationship. Ships with low Deadweight Density (e.g. Ferries) will tend to have high Emptyship Weight Fractions - as depicted in Kennell's Figure 7.20. This is true even for high speed ships (almost all of which are low deadweight density to date) - see the few added data spots in Figure 7.21 - and for aircraft (Figure 7.22.)

7.3.8 Conclusions on Kennell's Transport Factor

The Kennell TF formulation is another in a series of attempts to gain insight into the various niches of marine vehicle design choices. It obviously builds on the foundation laid by von Karman in 1950. In my opinion the single greatest insight attributable directly to TF is the realization that the

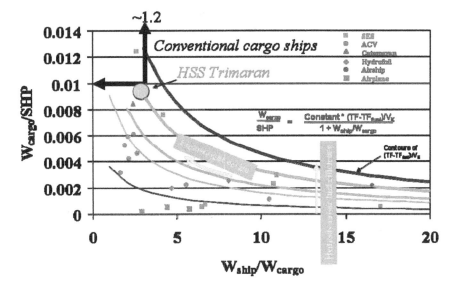

Figure 7.18: Kennell's finding of the relationship between ship weight, cargo weight, and SHP. Note here that W_{cargo}/SHP is in units of Long Tons per Horsepower. Note also that the W_{ship} on the x-axis refers to the weight of the empty ship, not the weight of the total ship.

lift weight may be decomposed into the various component weights of the ship, and that we may think of the TF "equation" as giving us a certain amount of TF to "spend" and then encouraging us to spend it wisely, to maximize that amount available for cargo, while recognizing the needs for TF(ship) and TF(fuel), and the items that drive them. In Kennell's terms, the TF shows us:

- There is structure to the universe

- High speed ships are different, but fit in with conventional ships

- Yardsticks/metrics are useful for establishing expectations

- Parametric assessments can provide useful insights

7.4 McKesson Parametrics

Von Karman invented the idea of transport efficiency, and its use to characterize the relative performance of various machines. Dr. Kennell developed

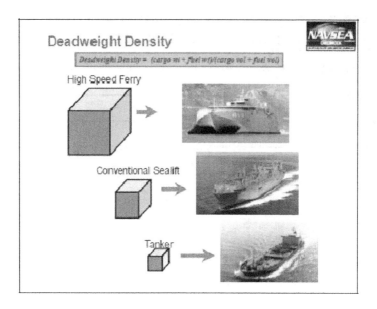

Figure 7.19: Kennell's graphic depiction of the nature of Deadweight Density for different ship types

the idea further with his "Transport Factor." McKesson has added to this body of discussion, with a five-parameter solution that can sometimes yield surprising insights into the domain of various ship types. This work was originally presented in 2006, [21] with the objective being to explore the major parameters that drive high speed military sealift vehicle design, and to use these parameters in a design mode to size the potential solution space for any given set of mission requirements.

Despite the focus upon the design nature of the task, McKesson's method is not a design tool. The solution space will include many different kinds of solutions, such as monohulls, SES, catamarans, etc, and the solution doesn't necessarily "know" which type of vehicle is being modeled. Neither does this technique tell the user the characteristics (length, for example) of the solution being modeled. It merely indicates that a solution is possible, of certain gross parameters.

Given this very top-level use for the tool, it is most emphatically not a tool for disputing fine-scale variations from one design to another. It may,

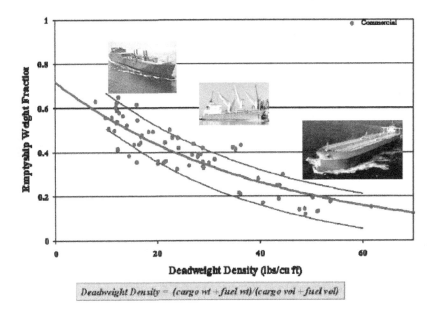

Figure 7.20: Low density payloads tend to demand higher values of lightship weight fraction

however, be a very effective "lie detector" at the top level.

7.4.1 The Sample Question

In 2005-2006 the US Navy's Office of Naval Research (ONR) contracted with Alion Science & Technology to assist in exploring the feasibility of high speed military sealift, under a program designated "HSSL." The question was intentionally left somewhat vague, so that researchers would enjoy the freedom to follow the most fruitful pathways. The requirements were also made deliberately demanding, in order to provoke innovation. The requirements were as follows:

- 5000 LT payload

- 43 knot speed

- 5000 nautical mile range

ONR's stated goal was to accomplish the above mission with a ship of less than 560 feet length and 12,000 tons displacement. McKesson's parametric

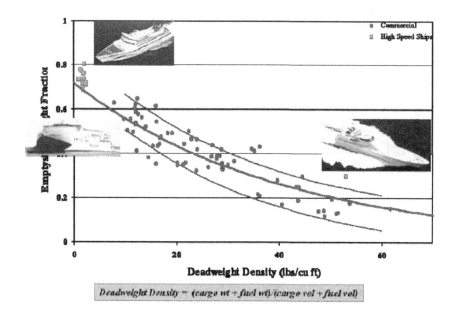

Figure 7.21: High speed ships follow the same trend

method was developed in order to find out if it was possible to perform this mission at the 12,000 ton displacement, or at least to measure how close to it one can get.

There is no shortage of concepts for high speed cargo ships. Instead, what is needed is a means for sorting through the myriad possibilities, and determining where the most fruitful avenues of exploration lie.

In this vein, this parametric methodology is also aimed specifically at helping to answer the question "Where must I make a breakthrough, in order to attain the HSSL desired level of performance?"

This section will introduce the method, and will explore some but not all of the possible ramifications and applications of the method. Some of these uses include:

- Use as a "lie detector" to detect claims that are well above the current state of the art

- Use as a predictive tool, to tell one where they will end up if they simply stick with the current state of the art

- Use as a thought-provoking tool, to nudge one toward the exploration

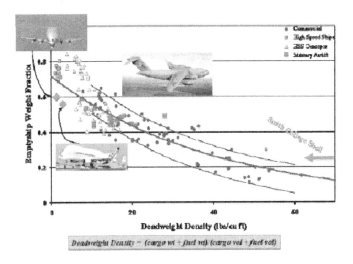

Figure 7.22: Even aircraft follow the same trend!

of concepts not normally entertained in naval architecture.

7.4.2 Major Parameters

McKesson finds that five major vehicle parameters govern high speed cargo carriage:

- Amount of power required, which depends upon:
    - Vehicle Lift / Drag ratio (to predict Drag)

- Ship weight: Fuel Weight, which depends upon
    - Overall Propulsive Coefficient (to convert Drag to Power)
    - Specific Fuel Consumption (to convert power to fuel weight)

- Ship weight: Light ship weight, which depends upon:
    - Weight of power
    - Weight of cargo carriage

143

In practice, the procedure flows as follows:
GIVEN: Payload, Range, Speed

1. Guess a Displacement.

2. From the TF trendlines, for that speed and displacement obtain a Power.

3. From that Power, obtain a weight for the Weight of Power.

4. From the Power, using the range and an assumed SFC, obtain the Fuel Weight.

5. From the assumed Displacement and an assumed k_{13456} factor obtain the weight of the remaining components of light ship weight (call this W_{13456})

6. Sum W_{13456}, W_{200}, W_{FUEL}, and W_{CARGO}.

7. Does the total equal the Assumed Displacement? If not, revise the Assumed Displacement and repeat until the values converge.

Each of these steps is expanded upon and explained, below.

7.4.3 Power Required - Ship Lift/Drag Ratios

Let us recall Dr. Kennell's TF, and the fact that it is reducible to simply Lift / Drag for the vehicles considered. In Figure 7.14 Kennell presented a curve of TF (or L/D) versus speed, for different ship sizes, and he (correctly) concluded that the L/D performance for a given speed varied with ship size.

But in Chapter 3 of this book, McKesson has reminded us that we can combine ship size and ship speed into one parameter - the Froude Number. What happens if, instead of using dimensional speed, Kennell had combined speed and size and plotted TF vs Froude Number? See Kennell's curve, reproduced again as Figure 7.23. Note that he provides trendlines for four trimarans at 100, 1,000, 10,000, and 50,000 LT. Let's consider just the last two of these, the 10,000 and 50,000 LT ships.

Both the 10,000 and 50,000 LT ships attain TF values of 40, but at different speeds. For the 10,000 LT ship a TF of 40 is attained near 30 knots, whereas for the 50,000 LT ship this occurs near 40 knots. This gives rise to Kennell's conclusion that TF depends on size. But note what happens if we look at the Froude Number. The volumetric Froude number for a 10,000 LT ship at 30 knots is 1.06. The volumetric Froude number for a

50,000 LT ship at 40 knots is 1.08 - virtually the same! Indeed, if Kennell's data is replotted against Froude Number instead of against dimensional speed we find that all of his lines collapse to a single curve.

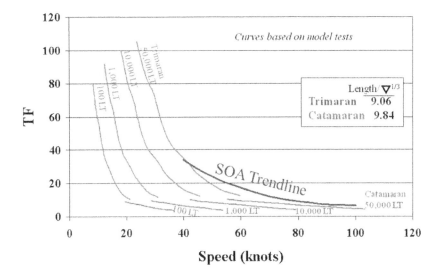

Figure 7.23: Kennell's curve showing the effect of size upon TF

7.4.4 McKesson "Observed Best Attainable" Curve

I analyzed my own data set, and produced a single curve of "observed best attainable" TF for real marine vehicles. Further, I produced an intentionally-simplistic curve fit to this "observed frontier." The curve and a few key data points are shown in Figure 7.24. What this yields is a very simple equation that may be used as a parametric predictor of "best practices" required ship power. I have intentionally kept the equation simple, both so that it can be easily remembered (the parameters are "five - four - three") but also so that it's very simplicity will serve as a cognitive reminder of the top-level nature of the analysis. Certainly one could generate a more precise fit of some sort, having values with five significant digits, but that is not the point of this technique, as will be demonstrated in subsequent paragraphs.

The point to be taken at this step is that we have a very simple equation, dependent only upon ship size and speed, that can yield a target value of propulsive power required, if one does a good job of designing the right ship for that point.

The "observed best attainable" curve is not a model of physics. It is, instead, an approximate description of the observed frontier or apparent state of the art. It does not state that a ship of $FN = x$ *must* have TF as given, but rather that it *could* have that TF, provided that the right choice is made for other parameters such as hull type, length, etc.

Further, TF is not a metric of ship "goodness." Instead, it is more accurate to think of it as an extremely simple ship resistance prediction formula.

Also, note that the Best Practices equation uses arguments that are surprisingly round numbers: 5, 4, and 3. This is intentional and serves two purposes: It results in an equation that is easy to remember, while at the same time the very roundness of the numbers reminds the user that this is not intended to be a high fidelity model, just a useful one.

One value of this TF curve is that it introduces the fact that resistance depends upon size. In 1997 in an earlier look at sealift I proposed a 40-knot L/D of 20, but as Figure 7.24 shows, it is easy to exceed that value - substantially - by making the vessel large enough. Indeed, according to the Best Practices Curve a TF of 100 is attainable at 43 knots, implying an L/D of say 130 or more, if the vessel displacement is approximately 700,000 tons. Unfortunately at this size, even with L/D=100 the required power would be over 3.5 Million horsepower. Clearly this latter is an absurd example, or at least one that lies outside the boundaries of the ONR HSSL project. However in a paragraph to follow I will return to more realistic explorations of the impacts of this dependency.

Finally, note that the "observed best attainable" curve is not a perfect fit of the data: There are some ships that exceed the curve. I note this, and will return to this in the later examples wherein I exploit these points.

This gives us a tool for the first of the five major parameters - TF as a Drag Predictor.

7.4.5 Weight of Power

By "Weight of Power" I mean the weight of the propulsion plant including engines and propulsors, but excluding fuel. For USN projects I use the weight of SWBS Group 200 for this item, and I refer to this factor as k_{200}, defined as $W_{200} = k_{200} * Power$. As a starting point let me assume a k_{200} value of 10 pounds per horsepower. That is to say that a 100,000 hp propulsion plant, including all of its components *including propulsors* may be expected to weigh about one million pounds, or ˜450 tons.

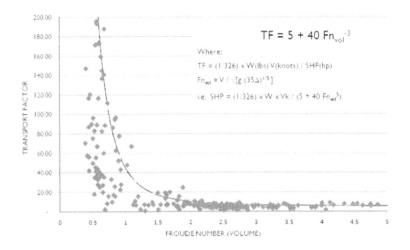

Figure 7.24: McKesson's Observed Frontier of ship TF.

A key observation on this step is that the weight of power depends only upon the choice of the propulsion plant type. I have opined that a high-speed diesel / waterjet power plant weighs approximately 10 lbs/hp. A diesel-electric / Azipod powerplant weighs approximately 60 lbs / hp. These values do not depend upon the power level selected, or any other parameter, although obviously one could use a parameter that does so depend, if additional precision were wanted.

By using this parameter, we are able to allocate a reasonable amount of weight to the powerplant whose size was determined from the TF curve. This weight, when added to all the other weights that make up the ship, must equal the displacement that we assumed to begin with. In TF terms we have allocated part of our TF_{total} to TF_{200}, with the remainder available for fuel, cargo, etc.

7.4.6 Fuel Weight

The next key parameter is the weight of fuel. This depends upon an additional parameter: the fuel efficiency of the powerplant.

The fuel efficiency of the powerplant, in the form of SFC (Specific Fuel Consumption) is the result of the engine type selection and powerplant design. SFC does not depend upon speed, nor upon range, nor upon level of installed power. Thus we can collect a menu or catalog of SFC values relevant to any number of machinery choices, ranging from say 150

147

grams / kiloWatt-hour for low-speed diesels, to say 400 grams/kW-h for gas turbines.

Note also that when we select the value of SFC to use, we must make sure that it accords with the value of k_{200} that we are using. Indeed, in reality what we select is the machinery technology, and for a given choice of machinery technology we look up the corresponding values of k_{200} and SFC.

For the HSSL project, because of the power levels that will be required, I looked only at gas turbine engines. Figure 7.25 shows the SFC reported for a variety of modern turbines in Navy service, plotted against their output power (Navy rating). Also included is a projection representing my estimate of what level of SFC performance might be attained by future larger engines - via a simple visual extension of the line.

However, when I plotted Figure 7.25 I was also aware that the engines plotted represented several different generations, and that the larger engines were generally newer. The same data is plotted in Figure 7.26 except here the ordinate is the year of introduction. Here again a shaded triangle is added guessing at what SFC's might be attained in the future.

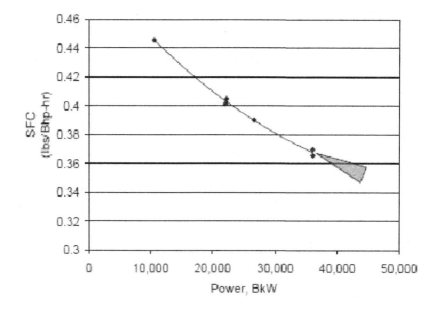

Figure 7.25: Propulsion Gas Turbine Engines, SFC versus Power, Current and Future Engines

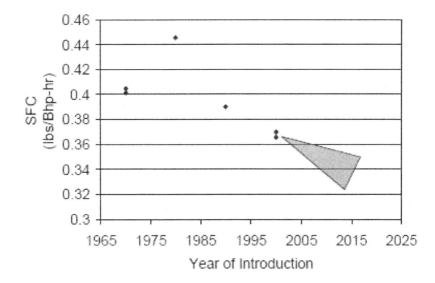

Figure 7.26: Propulsion Gas Turbine Engines, SFC versus Year of Introduction, Current and Future Engines

In consideration of this data, whereas the first investigation conducted below will use an SFC value of 0.40 lbs/hp-hr, it appears reasonable that SFC might be as low as 0.36 to 0.33 lbs/hp-hr.

7.4.7 Light Ship Weight

The final parameter needed to use McKesson's parametric method is one that controls light ship weight, other than the weight of power. That is to say, the weight of SWBS Groups 100, 300, 400, 500, 600, (& 700 if any.)

Consider first the weight of SWBS 100 - the ship structure: Reference [21] proposes the use of a Cargo Carriage Multiplier which is approximately the same as saying that W_{100} is proportional to W_{CARGO}. This reflects the fact that one part of the "job" of the ship structure is to enclose the cargo. Benford [22] earlier proposed to model structural weight using ship Cubic Number "CN", where: $CN = L * B * D/100$ He found that W_{100} was proportional to $CN^{0.90}$, and other lightship weight groups were proportional to $CN^{0.825}$. For purposes of a Very Simple Model, however, we can safely ignore these exponents and state that the weight is approximately linearly proportional to the pseudo-volume of "Cubic Number."

Cubic Number is the volume of a rectangular prism that encloses the entire ship. For ships of similar form, we may assume that this prismatic volume is some constant times the "real" total volume of the ship, meaning the sum of the underwater and above-water volumes. However, we may further choose to assume that the abovewater volume is some constant multiple (or fraction) of the underwater volume for a fixed mission (e.g. tanker, ferry, etc.) This then would collapse this volume formulation to state that total ship volume is directly proportional to underwater ship volume, where the constant of proportionality is determined by the ship mission.

Of course, this means that the W_{100} becomes some mission-dependent fraction of ship weight, such that we can replace the structural weight density with simply a structural weight fraction: $k_{100} = W_{100}/W_{TOTAL}$, a function of mission.

Moving now to SWBS 300, it is reasonable to assume that the electrical plant weight on a ship will depend approximately-linearly on the electrical load of the ship. The electrical load of the ship varies approximately as the ship volume [23]. Following the argument above about estimating this volume for W_{100}, this boils down to saying that the W_{300} fraction $k_{300} = W_{300}/W_{TOTAL}$ will be approximately constant for a ship of constant mission.

This same argument will then apply equally well to W_{500} and W_{600}, since again Redmond [23] suggests that these values are also scalable by volume, for a fixed mission.

In practice I group all of these SWBS groups together, resulting in a value I refer to as the "1356" factor, defined as follows:

$W_{1356} = W_{TOTAL} * k_{1356}$

This leaves us with W_{400} and W_{700} to estimate.

The weight of the SWBS 400 command and control suite of a ship is only weakly related to the size of the ship - radios and radars don't scale. I suggest taking the W_{400} weight as fixed across variations in a ship, or assuming a value for this weight if conducting a new design. Appropriate values may be reasonably guessed by a competent naval architect - from a few hundred pounds for a small craft to some tens of tons for a larger ship.

Finally we come to the shipboard weapon systems, SWBS 700. Shipboard weapon systems are actually the warship equivalent of "cargo" - they are the weight lift which constitute the ship's *raison d'être*. The present VSM therefore simply puts this into "cargo."

7.4.8 Putting it all together - A worked example

Consider now an example application of these simple parameters to the HSSL requirements. Recall that these were:

- 12000 LT Full Load

- 43 knots

- 5000 mile range

- 5000 LT Cargo

Figure 7.27 shows the following results: Assume a weight of 12,000 LT and a speed of 43 knots. This yields a Froude number of 1.48. The Observed Best Attainable curve suggests that we should be able to design a ship which, at this speed, will have an TF of 17.39. A TF of 17.39 with a displacement of 12,000 LT means an installed power of 204,000 hp.

At 10 lbs/hp this machinery suite will weigh 910 LT. If we assume an SFC of 0.4 lbs/hp-hr, then 5000 miles at 43 knots will require 4,200 LT of fuel. This fuel weight, plus the machinery weight, totals 5,100 LT. My first guess at k_{1356} is 0.33, based a small personal database for high-speed craft. This means that for the HSSL W_{1356} will be approximately 4,000 LT. Summing these all together, we find that we arrive at a ship capable of carrying 2900 LT of cargo, not the 5000 LT desired.

At this point we have quickly concluded that ONR's goal is not attainable, given the assumptions we have made. We could quit now, but this is premature given the assumptions that we have based our work upon. Let us instead use this insight to step forward and see which of those assumptions most needs to change, or where ONR most needs to make an R&D investment, in order to attain the desired performance.

7.4.9 A range of Examples

First, we say to ourselves "the displacement limit of 12000 LT was arbitrary. Let's see what happens if we increase the limit." Since the 12,000 LT ship carried 57% of the desired cargo, a linear assumption would lead us to expect that it will take a ship of 12,000/0.57 = 21,000 LT to carry the desired amount.

The second table (Figure 7.28) presents parameters for ten ships, where the displacement varies from the initial 12000 LT to 21,000 LT. We initially expect that, since the 12000 LT ship carried 2900 LT of payload, it will take a ship of about 21,000 LT to carry the desired 5000 LT of cargo. But we find that this is not the case, and that the goal value of cargo is attained at a much lower displacement (17,000 LT). In Kennell's words: "Size Matters."

7.4.10 The Design Space

What if the k_{1356} is other than 0.33? This parameter is probably the least defensible of my assumptions, so it makes sense to consider a fairly wide range of possible values. Similarly, I might also consider a variety of alternative propulsion technologies, trading off SFC against machinery weight (k_{200}.) Finally, we might, in the case of an ONR project, investigate what magnitude of hydrodynamic breakthrough would be required - by how much must we "beat" the Observed Best Attainable curve - to meet our goal?

Figure 7.29 depicts a curve wherein the k_{1356} varies from 0.25 to 0.65. A value of 0.25 is about right for a 70,000 tonne 20-knot container ship, and is probably optimistic for a high speed craft. A high speed ferry such as an INCAT catamaran has a k_{1356} value of about 0.65 (author's personal database.)

Note that in figure 7.29 the plotted value of displacement corresponds to a cargo weight of 5000 LT, with the design fully converged. We can see from this figure the sort of structural and auxiliary systems performance that would have to be obtained in order to meet, say, a 15,000 ton displacement target. By the same token, since full load displacement is a reasonable stand-in for cost, we can see from this curve the financial incentive to reduce the weight of structural and auxiliary systems.

At this point there are myriad further what-ifs that will have come to the mind of the reader. For example one can also generate a carpet plot against propulsion machinery technology, using two axes of machinery parameters of SFC and k_{200} (powerplant weight.) The point of this presentation is not to make a comprehensive analysis of the HSSL mission, but instead to show how performance metrics, starting with the foundation laid by Gabrielli and von Karman and built upon by Kennell, can be used to generate a useful high-level map of the ship design space.

Designers of AMVs are in the situation of Lewis and Clark, navigating without a map. It is nice to learn that we can draw our own reasonably useful map, based upon an admittedly simplistic model.

Assumed Displacement:	12000	LT
Required Speed:	43	knots
Required Range:	5000	n.miles
Required Cargo Weight:	5000	LT
Froude Number	1.48	[-]
TF (OBA)	17.39	[-]
Installed Power	203,940	hp
k_200 (assumed)	10	lbs/hp
W_200	910	LT
SFC (assumed)	0.4	lbs / hp-hr
Fuel Rate	81,576	lbs / hr
time en route	116	hrs
W_FUEL	4,235	LT
k_1356 (assumed)	0.33	
W_1356	3960	LT
W_1356	3,960	LT
W_200	910	LT
W_400	-	LT
W_700	-	LT
W_FUEL	4,235	LT
W_CARGO	2,895	LT
W_TOTAL	12,000	LT

Figure 7.27: A worked example of a Very Simple Model for the HSSL mission

Assumed Displacement:	LT	12000	13000	14000	15000	16000	17000	18000	19000	20000	21000
Required Speed:	knots	43	43	43	43	43	43	43	43	43	43
Required Range:	n.miles	5000	5000	5000	5000	5000	5000	5000	5000	5000	5000
Required Cargo Weight:	LT	3600	3600	3600	3600	3600	3600	3600	3600	3600	3600
Froude Number	[-]	1.48	1.46	1.44	1.42	1.41	1.39	1.38	1.37	1.36	1.35
TF (OBA)	[-]	17.39	17.89	18.38	18.85	19.30	19.74	20.17	20.58	20.99	21.38
Installed Power	hp	203,940	214,690	225,082	235,152	244,927	254,433	263,691	272,720	281,536	290,154
k_200 (assumed)	lbs/hp	10	10	10	10	10	10	10	10	10	10
W_200	LT	910	958	1,005	1,050	1,093	1,136	1,177	1,217	1,257	1,295
SFC (assumed)	lbs / hp-hr	0.4	0.4	0.4	0.4	0.4	0.4	0.4	0.4	0.4	0.4
Fuel Rate	lbs / hr	81,576	85,876	90,033	94,061	97,971	101,773	105,476	109,088	112,614	116,062
time en route	hrs	116	116	116	116	116	116	116	116	116	116
W_FUEL	LT	4,235	4,458	4,674	4,883	5,086	5,283	5,475	5,663	5,846	6,025
k_1356 (assumed)		0.33	0.33	0.33	0.33	0.33	0.33	0.33	0.33	0.33	0.33
W_1356	LT	4000	4333	4667	5000	5333	5667	6000	6333	6667	7000
W_1356	LT	4,000	4,333	4,667	5,000	5,333	5,667	6,000	6,333	6,667	7,000
W_200	LT	910	958	1,005	1,050	1,093	1,136	1,177	1,217	1,257	1,295
W_400	LT	-	-	-	-	-	-	-	-	-	-
W_700	LT	-	-	-	-	-	-	-	-	-	-
W_FUEL	LT	4,235	4,458	4,674	4,883	5,086	5,283	5,475	5,663	5,846	6,025
W_CARGO	LT	2,855	3,250	3,655	4,068	4,488	4,914	5,348	5,786	6,231	6,680
W_TOTAL	LT	12,000	13,000	14,000	15,000	16,000	17,000	18,000	19,000	20,000	21,000

Figure 7.28: How much must the displacement grow, to obtain the targeted value of cargo?

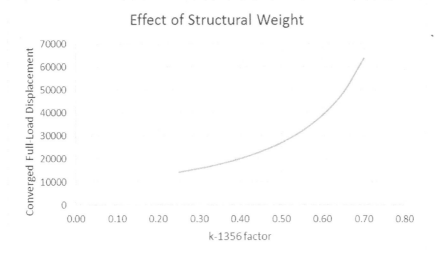

Figure 7.29: Map of Ship Size Versus k_{1356}. This shows the impact of specifying heavy solutions for structure and auxiliary systems, or
contrariwise the incentive for developing lightweight structure
and auxiliary systems. (Corresponds to 5000 LT cargo, 43
kts, 5000 nmi range, TF per Observed Best Attainable Curve,
Weight of Power = 10 lbs / shp.)

—

8 Hydrostatic Balance

For all the other courses in naval architecture, the hydrostatic balance relationship is very simple: Weight = Buoyancy. Mathematically this is written:

$$W = \rho g Vol \tag{8.1}$$

For the dynamically supported vehicles we have to add the weight borne by the dynamic lift, thus:

$$W_{dynamic} = C_L \frac{1}{2} \rho S V^2 \tag{8.2}$$

For the air cushion vehicles we have to redefine the volume in the static case, to account for the presence of the air cushion. The cushion is nothing more than a displacement, where pressure can be substituted for draft, as follows:

$$Vol_{cushion} = Area(cushion) * Pressure(cushion)/(\rho_{water} g) \tag{8.3}$$

Area times Pressure is of course Force. To express this as an equivalent displacement volume we divide by the density of the fluid being displaced - the fluid upon which the vehicle is floating: Water.

This approach treats the air cushion as a displaced volume, exactly like a box barge with no bottom - a barge whose bottom is formed only by the constant-pressure boundary condition.

We can then combine these parts to describe all three types of sustention, thus:

$$W = \rho g Vol_{hulls} + C_L \frac{1}{2} \rho S V^2 + P_c A_c \tag{8.4}$$

Where P_c is the cushion pressure and A_c is the cushion area.

Note that in this formulation the cushion is modeled as being wall-sided. It is imaginable that one might have a hull shape such that this assumption is not correct, because the cusion area "masks" some of the hull volume. It is important to model this carefully so that one does not double count any fraction of the lift.

For most air cushion craft the cushion is rectangular in planform shape, or nearly so, so the cushion area may be written as $A_c = L_c * B_c$. Continuing to play with the cushion volume relationship we see that the cushion-borne weight, W_c, is $W_c = P_c * L_c * B_c$. This describes a floating rectangle, having a block coefficient of 1.0, with a draft of "P_c" (more properly, the draft is $P_c/(\rho_{water}g)$).

We also, in ACV and SES design, encounter a metric of P_c/L (pronounced "Pc upon L") which is a measure of cushion pressure or cushion density. You may also think of it as being similar to a Draft-to-Length ratio. Since an SES has a rectangular planform, it has a very blunt entry in the waterplane, but its entry in the buttock direction is P_c/L, and thus I encourage thinking of P_c/L as the ACV equivalent to a Length-to-Beam ratio. Typical values of P_c/L are in the range 1%-2%.

In an SES we have, in addition to the air cushion, some perhaps 20% of the sustention borne by sidehull hydrostatics. These hulls are of course governed by all the same concerns as any catamaran hull. We should also note that, given the high speed of SES, these sidehulls may generate important hydrodynamic forces, including some dynamic lift.

Finally, let us note that the dynamic lift component, the C_L, varies at least with trim angle and that the total dynamic lift varies with speed squared. The weight of the craft probably doesn't change with speed, so in the case of dynamic lift craft there will clearly be a change in the hydrostatic volume as a function of speed. And in fact this is why planing craft go fast: Because they unload the hydrostatic component of lift, and in so doing they reduce the hull wetted surface and other factors which contribute to drag.

And that leads us into our lectures on resistance.

9 SWBS 051 - Resistance

9.1 The Resistance Components

For the first estimate at resistance I like to come up with a "Target" R_T, rather than an estimate of my ship's R_T. For this purpose I use the Observed Best Attainable curve given earlier of: (see Chapter 7.)

$$TF_{OBA} = \frac{W * V}{Ps} = 5 + 40 * Fn_{vol}^{-3} \qquad (9.1)$$

If I apply an assumed value for overall propulsive efficiency, then this yields a goal-value for resistance, a drag value I hope to attain or beat.

The components of resistance for a low-speed monohull, according to Froude's formulation, can be written as wavemaking and friction. Some authors will add a small amount for appendages and windage. Thus, in the traditional method:

$$R_t = R_f + R_r + R_{air} + R_{correlation} \qquad (9.2)$$

Where:

- R_f - Frictional resistance is found from flat-plate methods

- R_r - Residuary resistance is found from model tests

- R_{air} - Aerodynamic resistance is estimated by application of an air drag coefficient to the frontal area. For large slow ships this term is such a small part of the total drag that it is often even ignored. Obviously, however, it increases rapidly with speed. Further, for ships with low hydrodynamic resistance this aerodynamic resistance comes to form an even larger part of the whole. It is important therefore not to ignore it. That said, however, it is subjected to normal treatment and does not get a lecture in this course. See Hoerner and other similar sources to come up with reasonable air drag coefficients for the vessel of interest.

- $R_{correlation}$ is a correlation factor which is intended to account for scale effects, particularly those on the frictional drag (also known as ΔC_f or C_a.)

By contrast, the Advanced Marine Vehicle has resistance components as follows:

$$R_t = R_f + NR_r + R_{inter} + nR_{cushion} + R_{skirt} + R_{spray} + R_{append} + R_{air} + R_{waves}$$
$$(9.3)$$

That is to say:

- Friction

- The residuary or wavemaking resistance of 'N' hulls

- Resistance caused by interference between the hulls

- Resistance caused by 'n' air cushions (if present)

- Resistance caused by air cushion skirt systems (which may itself be broken down into frictional and residuary components)

- Resistance due to spray generated by the hull

- Resistance due to appendages (which is no longer small, due to the high speed of the ship)

- Resistance due to windage (which is no longer small, due to the high speed of the ship)

- Resistance due to encountering ocean waves

Can we estimate all of these? Let's take a second look at them:

R_f = Frictional drag - as normal except that wetted surface may vary tremendously depending on parameters such as cushion pressure and speed.

R_r = Residuary resistance. Usually we treat this as the residuary resistance of the hulls, although in a model test program it will pick up bits and pieces not accounted for elsewhere, which can cause problems.

R_{air} = aerodynamic resistance. While this is small for conventional ships, at the speeds that AMVs work at this can become a substantial factor.

$R_{correlation}$ = The correlation allowance C_a is rarely discussed in fast ship literature, but there is a general agreement that it takes on a large importance in this arena. This is because the traditionally-important factors such as R_r have been reduced so much that the magnitude of C_a becomes relatively large. Unfortunately there is really no agreement as to what to do about this, and we won't teach on it in this course. Actual practice seems to vary from one tow-tank to another. In my experience I have mostly seen Ca values of zero.

$R_{interference}$ = Interference effects between multiple hulls - can be included into Residuary Resistance, subject to some cautions discussed below.

$R_{cushion}$ = Wavemaking drag of air cushion - should be calculated independently

$R_{momentum}$ = Lift System Air Momentum Drag - should be calculated independently

R_{spray} = Spray and Spray Rail Drag - According to Faltinsen can be 12% of the total resistance of the craft [4], but nevertheless will not be treated in this course. Gets lumped into Rr.

$SkirtDrag$ - not clear if this is a residuary component or a frictional component, both terms should be used - see [24].

$AppendageDrag$ - Another component that will not be treated in this course. In general the most effective AMVs recognize the importance of appendage drag by simply avoiding appendages altogether, as far as possible. This takes the form of using waterjets instead of rudders, etc. This philosophy notwithstanding we do commonly encounter ride control devices, which may include Foils, Interceptors, and Wedges or Tabs. These, of course, have drag. We may also encounter various ancillary devices such as high speed rudders, or even the rather astonishing suite of appendages used on the SES 100A for stability enhancement, see Figure 12.18. Nevertheless, we will assume that the resistance of these devices is understood and derivable from "conventional" naval architectural practice, and we do not treat of it here. Note that under this heading we would include Hydrofoil Drag (Drag due to lift, incl tip effects)

Faltinsen [4] gives some interesting figures in his Figures 4.1 & 4.2 showing the relative importance of the various drag components, as a function of speed, for a catamaran and an SES. These figures are reproduced as Figures 9.1 and 9.2 here.

Also, Doctors has shown that the various innovations of some of the AMV hull types, such as the use of slenderness or wave cancellation, has the effect of reducing the wavemaking drag to such a degree that the frictional drag attains great importance in the overall craft design balance. This remark is an interesting counterpoint to the rule of thumb used by some warship designers that the frictional and residuary drag components of a well-designed hull should be in 1:1 balance, i.e. approximately equal. Faltinsen's curves would seem to support this rule of thumb - in Figure 9.1 we see that friction accounts for 50% of the calm-water drag.

Finally, since it doesn't seem to fit neatly anywhere else, I insert an observation by Dr. Larry Doctors (2008 - verbal) that the Doctors & McKesson FAST 05 ([25]) results clearly show that from the perspective of wave drag it doesn't matter whether you have a cushion or not. Unlike the effect of

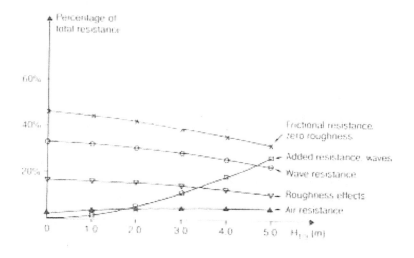

Figure 9.1: Drag components of a 70m catamaran, from Faltinsen [4]

slenderness, the effect of the cushion is not to reduce the wave drag of the ship, but rather to reduce the frictional drag. This is also the reason that the Keck Sea Train works so well, because it permits a sidehull form that is the minimum required to contain the bubble, and thus has the minimum possible wetted surface, coupled with a very slender overall ship form.

In the units which follow, we will consider each of these drag components, and I will provide advice on how they may be estimated practically.

9.2 Frictional Resistance

Pause and consider the implications of the Froude method of extrapolating resistance: We measure the total resistance of a model. We know that the frictional and residuary components do not scale in the same way, so we must separate them. We assume that the friction of the model is the same as the frictional drag of a flat plank of the same area. In effect we are saying "Our knowledge of friction is so good that we can accurately calculate it. In model tests we calculate R_f, and we can accurately subtract it from R_{total}, with confidence that the remainder is $R_{wavemaking}$."

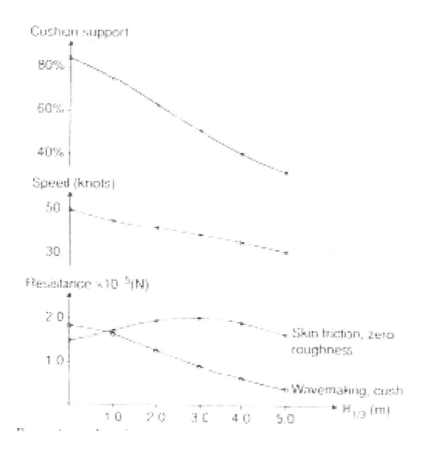

Figure 9.2: Drag components of a 40m SES, from Faltinsen [4]

In practice we do this by applying the familiar equation:

$$R_f = \frac{1}{2}\rho SV^2 C_f \qquad (9.4)$$

Where: $C_f = f(Rn)$, and we take this from one of several C_f curves.

But which C_f curve shall we use? Consider Faltinsen's comparison of five different C_f formulae, reproduced in Figure 9.3. The three numbered equations he uses are:

- ITTC: $C_f = .075/(\log_{10} Rn)^2 - 2$

163

- Eq 2.66: $C_f = 0.0303 Rn^{-1/7}$

- Eq 2.67: $C_f = 0.066/(log_{10} Rn - 2.03)^2$

What Faltinsen shows is that C_f may be in error by 10-15%. And this is the component of resistance that we claim to know "so well!'

Further, what about roughness? The C_f curves are for smooth flat plates. In "regular ship" design we worry about the "flatness" assumption, hence the use of a Form Factor $(Rf = (1 + k)Rf_{flat})$. It may be argued that in AMV design we're closer to flat because of our slenderness. But are we "smooth?" Faltinsen says no, and recommends use of a friction formulation with explicit modeling of roughness effects [4].

And finally, even so simple a measure as the Reynold's number is fraught with unexpected uncertainty. Reynold's number depends upon viscosity. Viscosity varies with water properties and temperature. This means that, in principle, lake service should be different than ocean service, and tropical service should be different than temperate service.

We find that, once again, the AMV designer is trying to perform a precise optimization with imprecise tools. Continuing my Lewis & Clark metaphor, it is like trying to navigate the area around Louisiana using maps like the one reproduced in Figure 9.4. It can be done, but one would be prudent to be not too trusting.

Rn	C_F, "Exact" (White 1974, Table 6.6)	C_F, ITTC	Error, %	C_F, eq 2.66	Error, %	C_F, (Hughes, eq 2.67)	Error, %
10^6	0.004344	0.004688	7.9	0.004210	-3.1	0.004188	-3.6
10^7	0.003015	0.003000	0.5	0.003030	0.5	0.002672	-11.4
10^8	0.002169	0.002083	-3.9	0.002181	0.5	0.001852	-14.6
10^9	0.001612	0.001531	-5.0	0.001569	-2.6	0.001359	-15.7
10^{10}	0.001236	0.001172	-5.2	0.001129	-8.7	0.001039	-15.9

Figure 9.3: C_f Curve Comparison, from Faltinsen [4]

But Lewis & Clark had uncertainties too - and plenty of them! My goal is not to make us throw up our hands in despair, nor to over dramatize the situation: The ITTC curve and Standard Seawater have worked as useful fictions for more than my lifetime. I do not propose overthrowing them, I merely wish to highlight the uncertainties that we live with.

In this course we shall use the ITTC 1957 C_f line, with no roughness allowance, and standard 15C seawater.

Figure 9.4: AMV design often feels like navigating using maps like this

9.2.1 Wetted Surface Variation

Having chosen our flat-plate friction line, the next data that we need is the ship's wetted surface. In conventional ship practice we determine a single value of wetted surface at each displacement. In AMV design we have the added complication that wetted surface is a function of speed. This is because not only does the craft rise up on various planes with speed, but also her own generated waves will affect her wetted surface. This will occur not only outside the hulls but also in between the hulls of a multihull. In the case of an SES the wetted surface will also depend upon the cushion pressure.

How large is this effect? How much does the wetted surface vary from at-speed to at-rest conditions? Let's discuss it, and look at some pictures and data:

We shall ignore the form effect upon viscous drag, and we shall pretend that all of the viscous resistance can be adequately modeled by the application of a frictional resistance coefficient such as $C_f = .075/(\log Rn - 2)^2$. As *practical* AMV designers, the key novelty here is that the wetted surface of the AMV can vary - tremendously - with speed. Therefore it is necessary to use a friction formula wherein wetted surface is a function of speed - $WS = f(V_k)$.

In most cases this is accomplished by model testing. During model testing the model must be equipped with means of determining the dynamic wetted surface, and this dynamic area is used during the Froude extrapolation of the test results. If you have relevant parent data you may be able to use the parent data $WS_{dynamic}/WS_{static}$ relationship to estimate the dynamic wetted surface of the offspring craft.

Before model tests it would be nice to have a predictive method for estimating the dynamic wetted surface. Firstly, let us note that we don't have any cookbook ways to predict this for displacement hulls. Model tests work OK, but how about before model tests?

Some of the variation in wetted surface is due to dynamic sinkage and trim, so we can attempt to see how big these values are and derive wetted surface implications from them. Figure 9.5 shows one set of sinkage data from [4], and one might logically create a relationship between sinkage and wetted surface from this result.

Some of the variation is due to own-ship waves, and we can use CFD to predict these. Depending upon the degree of control we have over the CFD program, we may be able to get it to report the dynamic wetted surface directly.

For SES one can use Kolazaev's method, given by Yun & Bliault [8] as:

$$S_f(Fn) = K_f(Fn)S_{f0} \qquad (9.5)$$

where:

- $S_{f0} =$ Calm water wetted surface

- $Fn =$ Froude number on Cushion Length

- $K_f(Fn)$ given by Figure 9.6:

Once model tests are engaged the situation becomes a little easier. Model tests with photographs are one accurate way to measure wetted surface. I have also had acceptable results using girth-measuring tapes fitted to a model at two or three stations longitudinally. Figure 9.7 depicts the Alion HSSL model with wetting tapes fitted. This craft is an extremely high L/B SES, and as such may have very little dynamic change in wetted surface, but it is the only such photograph in the author's library. Figure 9.8 shows the speed-variation in wetted surface measured by this method.

McKesson's preferred technique for SES is as follows:

1. Assume outside wetting based on normal hydrostatic calculations.

2. Assume inside static wetting based on the cushion depression being flat

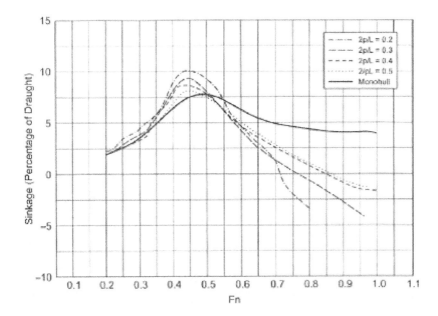

Figure 9.5: A reproduction of Faltinsen's reference on Running Sinkage of a catamaran [4]

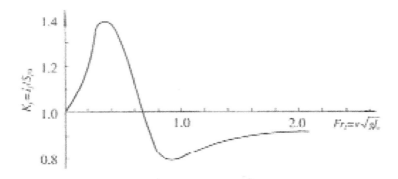

Figure 9.6: Kolazaev's figure for $Kf(Fn)$

3. Measure inside and outside wetting via resistance tapes in model tests.

4. Develop SEPARATE curves of $S_{dynamic}/S_{static}$ for inside and out-

Figure 9.7: The wetting tapes (the two gold strips) fitted to the HSSL model to measure wetted girth. Three such sets of tapes were installed at different stations along the length of the model.

side cases. Curves should depend on Froude number, but it will be FN_{LWL} for the outside case, and $FN_{cushion}$ for the inside case.

9.3 Wavemaking (Hull, not Cushion)

Having established techniques for estimating the frictional drag of our AMV, let us now turn our hand to estimating the wavemaking or residuary component of drag.

En passant let me mention that I understand, for dynamically supported craft, that the wavemaking drag to weight ratio (inverse L/D ratio) is uniquely related to the ship's dynamic trim, as $R_w/W = \tan(trim)$. Perhaps it is inappropriate to admit this in a textbook, but I do not fully understand this truism and look forward to learning more about it. I do know however that very practical use can be made of this relationship: I have measured the dynamic trim (dynamic trim is the change in trim with

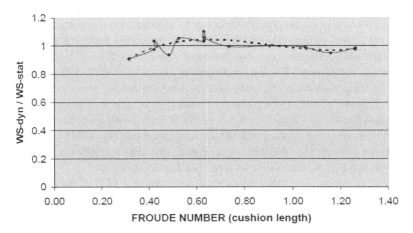

Figure 9.8: The dynamic wetted surface variation with speed as measured on the HSSL model

speed, the difference between the at-speed trim angle and the static trim angle) using long pendulums on a high speed ferry, and by plotting the tan($trim$) data obtained a very clear depiction of where the humps and hollows in the ship's wavemaking drag curve lay. Alternatively, one should be able to take predicted wavemaking drag data and invert the relationship so as to predict the dynamic trim.

9.3.1 Estimating Wavemaking Drag of a Single Slender Hull

Fortunately, the hulls of the vast majority of AMVs are truly slender, and all the hydrodynamic simplifications that go under the name of "slender ship theory" can be applied with excellent results in very practical cases. I will present techniques that rely on the following methodologies:

- Computational predictive methods

- Systematic Series hull predictions

- One-Off parents (Worm Curves)

- Model extrapolations

169

Computational predictive methods

When there is no relevant parent hull nor systematic series to draw upon, and it is too early in the design process to conduct model tests, we have to rely on purely computational methods for resistance prediction. Of the computational methods there are two that rise to the fore. The first, and most easily dispatched in this textbook is the use of CFD. There are a variety of CFD tools that are quite mature, and improving almost daily. For that component of the AMV resistance task that is simply the wavemaking resistance of a single slender hull, these CFD tools work quite well.

In a few later paragraphs I will tell a cautionary tale of a CFD prediction of interference drag that did not go very well, but that is a different element of the resistance problem than we are treating at this time.

CFD is, of course, the attempt to solve some version of the Navier Stokes equations explicitly. In consequence, CFD has its roots as far back as the very development of those equations (which was around 1800.) The challenge is that the deceptively simple equations are extremely difficult to solve.

A more simple equation was developed by J. H. Michell in 1898, and is called Michell's Integral. Michell's integral may be derived as follows: (Thanks to Leo Lazauskas for this presentation.)

The disturbance velocity potential can be generated by a distribution of Havelock sources over a region R of the hull center-plane $y = 0$ with strength $m(\xi, \zeta)$ per unit area at the point (ξ, ζ). Thus,

$$\phi(x, y, z) = \iint_R d\xi d\zeta \, m(\xi, \zeta) \, G(x - \xi, y, z; \zeta) \tag{9.6}$$

It is assumed that $z \geq \zeta$ over the vertical plane R, which is always true on the free surface $z = 0$.

According to Michell's thin-ship theory, the source strengths, $m(\xi, \zeta)$, in equation (9.6) are proportional to the longitudinal slope, namely

$$m(\xi, \zeta) = 2U Y_\xi(\xi, \zeta). \tag{9.7}$$

On substitution into equation(9.6), we get

$$\phi(x, y, z) = \frac{U}{2\pi^2} \Re i \int_{-\pi/2}^{\pi/2} d\theta \int_0^\infty dk \, k_1 e^{-ik\varpi} [e^{-kz} - K_1(k, \theta) \, e^{kz}](P + iQ) \tag{9.8}$$

where

$$P + iQ = -\frac{1}{ik_1} \iint_R d\xi d\zeta \, Y_\xi(\xi, \zeta) \, e^{ik_1 \xi + k\zeta} \tag{9.9}$$

and where $k_1 = k \cos \theta$ is the x-wise wave number.

Integrating equation (9.9) by parts yields an expression for $P + iQ$ that depends on the actual hull offsets, rather than on the hull slopes. Thus,

$$P + iQ = \iint_R d\xi d\zeta \, Y(\xi, \zeta) \, e^{ik_1\xi + k\zeta} - \frac{1}{ik_1} e^{ik_1\xi_e} \int_{-T}^{0} Y(\xi_e, \zeta) \, e^{k\zeta} d\zeta \quad (9.10)$$

where ξ_e is the ξ-ordinate of the stern.

The result of the integration by parts given in equation(9.10) assumes that all offsets at the bow are zero, i.e. that $Y(\xi_b, \zeta) = 0$, where ξ_b is the ξ-ordinate of the bow. The second term of equation(9.10) vanishes if there is no transom stern or if there is no additional offset due to the inclusion of the BL displacement thickness.

The linearised wave elevation $z = Z(x, y)$ is given by

$$Z(x, y) = -\frac{U}{g} \phi_x(x, y, 0) \quad (9.11)$$

which is, in effect, a quadruple integral given by

$$Z(x, y) = \frac{1}{\pi^2} \int_{-\pi/2}^{\pi/2} d\theta \int_0^\infty dk \, e^{-ik\varpi} K_2(k, \theta) \, (P + iQ) \quad (9.12)$$

where

$$K_2(k, \theta) = \frac{k^2}{k - k_2}. \quad (9.13)$$

Far behind the vessel, i.e. when x is large and positive, the major contribution to the wave elevation comes from the residue at the pole $k = k_2$. The far-field wave elevation is then given by the "Linearised Ship-Wave Intgeral",

$$Z_F(x, y) = \Re \int_{-\pi/2}^{\pi/2} A(\theta) \exp[-ik_2\varpi] \, d\theta \quad (9.14)$$

where A is the complex amplitude, or "free-wave spectrum" given by

$$A(\theta) = -\frac{2i}{\pi} k_2^2 (P + iQ). \quad (9.15)$$

The total energy in the far-field wave pattern yields Michell's integral for the wave resistance.

Fortunately for the journeyman practical AMV designer, there exists freeware software which uses Michell's integral to estimate the wave resistance of a slender hull. This software, developed in Australia by Leo Lazauskas above quoted, is called "michlet." It is available as a free download from Leo's website at www.cyberiad.net and is a practical tool for estimating the wavemaking drag of slender AMV hulls.

Leo Lazausakas and his colleague Ernie Tuck have published many very interesting papers based on the exploitation of Michell's integral (e.g. References [5] & [26].) A few sample outputs are presented in Figure 9.9, which depict both the wave pattern and the resulting wave resistance for a single hull in deep water, at different Froude numbers.

Series hull predictions

There are several useful systematic series results with slenderness ratios of interest to the AMV designer. I find the following to be particularly useful, although I am sure there are others:

- The Taylor Standard Series (Reference [27]) is not ridiculous for some applications.

- Series 64 (Reference [28]) is useful for Trimaran Amas.

- Lundgren & Williams' SSPA series (Reference [6]) is useful - and easy to use.

- VWS89 Catamaran Series (Reference [29])

Most of these series are commonly available in mainstream naval architecture texts and I shall not repeat those explanations here. Also note that many of them are codified in software such as NAVCAD. In 1996 Molland et al (Reference [30]) presented results of systematic series tests of slender hulls, i.e. catamaran demi-hulls or trimaran center hulls. They found - among other conclusions - that the displacement length ratio (Δ/L^3) was the most important hull parameter, dominating the effect of secondary parameters such as block coefficient or B/T ratio, etc. This is quite a useful result, as it tells us which parameter is most important to 'get right' when selecting a systematic series (or a parent hull, for that matter.)

I particularly like the Lundgren SSPA series. I find that many AMV hulls lie within it's range. I am surprised to find it absent from tools such as NAVCAD, so I will mention it further here. Figure 9.10 from the Lundgren paper presents the range of applicability of the 'present series' as compared to other well-known series. Note that the two columns 'FnL' and 'B/T' have their labels switched.

The Lundgren series provides easy-to-use curves of C_R versus Froude Number, for discrete values of B/T and Slenderness. One such data set is presented in Figure 9.11. The user of this data need only perform interpolations to arrive at his target values of B/T and $L/\bigtriangledown^{1/3}$, and generate a C_R vs FN curve therefrom. This curve can then be used as the predictor of wavemaking drag for the hull in question, and the other components (interference of multihulls, etc.) can be added *post hoc.*

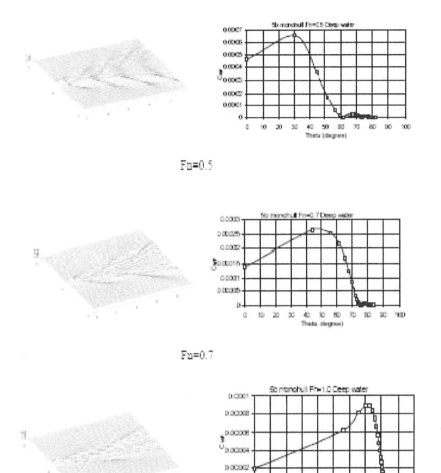

Figure 9.9: Wave pattern and distribution of wave pattern resistance as estimated by Michell's integral, from Lazauskas and Tuck [5]

Model extrapolations

Of course, model tests are excellent tools for predicting the resistance of slender hulls, and of AMVs *in toto*. There are, however, a few notes and

Model family	C_B	Range of $L/\nabla^{1/3}$	B/T	FN_L
Ext. Taylor Series	0.50 – 0.68	7.6 – 10.6	3 – 4.5	< 0.9
Series 62	0.44 – 0.50	4.1 – 7.7	4.2 – 7.8	< 3.0
Series 64	0.35 – 0.55	8 – 12.4	2 – 4	< 1.5
KTH/NSMB Series	0.35 – 0.55	5.8 – 7.8	3.2 – 4.4	< 0.9
EMB Series 50	0.35 – 0.42	5.5 – 9	4 – 15	< 1.9
Present Series	0.40 – 0.55	6 - 8	3 – 4	< 1.3

Figure 9.10: Lundgren SSPA series parameters compared to other series

quirks which apply. Firstly, let us recall the fundamental relationship that we use when extrapolating model test data: $C_R = R_R/[\frac{1}{2}\rho S V^2]$. Remember the discussion above about the uncertainties in estimating the wetted surface, and the fact that the wetted surface changes with speed and other parameters? This means that our derived value of C_R will suffer from uncertainty in the same degree. One can work around this, by being extremely precise and making sure that the definition of S is managed carefully, i.e. that the same dynamic correction factor is used for the model and for the ship. But I think that there is a simpler and in fact better solution:

I claim that we know Δ much better than we know 'S' since weight doesn't vary with speed. I therefore recommend that such extrapolations should be performed on the basis of Rr/Δ e.g:

$$Rr_s(V) = Rr_m(V)\frac{\Delta_s}{\Delta_m} \tag{9.16}$$

Or first develop a performance curve for the model:

$$K_r(V) = \frac{Rr_m(V)}{\Delta_m} \tag{9.17}$$

Followed by an extrapolation to full-scale for the offspring:

$$Rr_s(V) = \Delta_s K_r(V) \tag{9.18}$$

Note that by the notation $Rr_m(V)$ I mean the value of Rr_m at a particular value of V.

One-Off parents (Worm Curves)

Frequently we have a case where we are developing a ship that is similar to some previous ship, but not an exact geosim. In this case it is very helpful to perform a resistance estimate using the previous ship as a single-case parent, developing what is called a "worm curve" against some other systematic series. This is not an AMV-specific technique, but since it may not be well known it is worth explaining here.

The Worm Curve method is a technique for using systematic series data to model the variation of a hull form that isn't a member of that series. It is commonly used with warships, wherein parent-ship data is compared to an equivalent Taylor Standard Series hull. In effect one is saying 'NewShip will differ from Taylor Series the same amount that OldShip differs from Taylor Series.' This is a little clearer when written mathematically:

$$WCF(Fn) = \frac{R_r(Fn)_{parent}}{R_r(Fn)TSS_{parent}} \tag{9.19}$$

$$R_r(Fn)_{newship} = WCF(Fn)R_r(Fn)TSS_{newship} \tag{9.20}$$

Steve Toby wrote the following description of the Worm Curve, as Appendix A to his 2002 paper "To the edge of the possible." [31] Steve's description is specific to the use of a Taylor Series Parent, but I believe it is applicable to use of parent forms from other series as well - always requiring, of course, that one sticks with a fixed parent series once the worm curve is calculated (i.e. don't use a Taylor worm curve on Series 60 data!)

Toby: *"To estimate the resistance of a ship of present day form, the equivalent Taylor series hull can be run to find the residuary resistance. This equivalent hull has the same Taylor series parameters, however it does not necessarily have the same dimensions. A ship of the same dimensions and prismatic would usually have the wrong displacement because of the difference in Cx. Therefore the equivalent ship could have the same length, displacement, Cp, and beam/draft ratio, but beam and draft will not be the same. Other combinations could also define an equivalent ship.*

Now the residuary resistance needs to be adjusted for the difference in hull form between the Taylor equivalent ship and the actual hull form. This is done using a number called the worm curve factor or WCF. The WCF is a ratio between the residuary resistance of our hull and that of the equivalent Taylor hull. It is often expressed as a ratio of residuary resistance per ton of the two hulls, because if the equivalent Taylor hull is of different displacement, the residuary resistance cannot be used. The WCF is calculated as a function of SLR [Speed-to-Length Ratio] from a model test of a hull of form similar to the hull being analyzed. An unspoken presumption is that the WCF is still valid for hulls of different proportions if they are of similar shape. A WCF is computed from a model test automatically using the Taylor series computer program in current practice.

Now let's revisit some of the assumptions behind the use of a WCF. The key assumption behind it all is that the WCF represents hull form only and is not influenced by proportions - that is, the Taylor variables can be separated out from the hull form variables covered by the WCF. This is 'mostly' true, but there are some inaccuracies. Research performed for Part 1 of this paper indicated that plotting residuary resistance against beam to draft ratio resulted in a nonlinear curve at some speeds but not at others (in the Gertler data and in the TSS86 computer program, not in the original Taylor data). This in turn suggested that a favorable WCF value didn't mean as much at some B/T ratios as it did at others, and therefore, that there is some degree of nonlinearity or coupling between the WCF and the Taylor variables. Likewise, recent experience shows that destroyers with high displacement-length ratios often show favorable WCFs, as if the Taylor series exaggerates the penalty of DLR (Displacement-to-Length Ratio) slightly. Therefore, we should view even the most basic assumption behind WCFs with caution; in fact, it is not certain that hull form and the Taylor variables are completely uncoupled."

"..."

"Finally, there is one unspoken, non-technical assumption in the use of worm curves that is very likely to trip up the unwary. This is the fact that the worm curve is not a measure of merit. When the WCF is plotted it is easy to fall into the trap that if it is well below Taylor at the design speed, the ship is a 'good' ship, while if it is above Taylor, it is a 'bad' ship. While this is sometimes true, it is not always. There are three reasons why it may not be true:

1. *The residuary resistance, whose behavior is represented by the WCF, is not all of the resistance. Therefore, if you had to add lots of wetted surface to get a good WCF, your ship could be a 'dog' even if its WCF was below 0.8.*

2. *At low speeds, residuary resistance may be so small a percentage of the total that a WCF of two will only add a few percent to the total resistance of the ship.*

3. *It makes a lot more difference if you picked the right proportions than if your WCF is good. For example, take a ship of Fletcher dimensions; its WCF at 35 knots is .84, and its resistance is 306,200 lb. Stretching the Fletcher lines out to a DLR of 40 (the optimum according to the 1997 destroyer paper) by increasing length to 415', and keeping*

displacement constant, results in a resistance at the same speed that's about 10% less even with a WCF of one for the long hull.

Therefore, using the WCF as a measure of merit should be avoided."

On the wavemaking resistance of SES sidehulls

The preceding discussions have all focused on round-bilged type hulls, as may be found on catamarans or trimarans. These techniques can and have been used for SES as well, but SES also admit of some additional techniques that are worthy of mention here.

SES sidehulls lend themselves to two separate methods of treatment. I tend to use the first one in early design stages, and then shift to the second method as the design matures.

In the first method (ref [8] page 111) we simply enlarge the beam of the cushion to an 'equivalent beam' to account for the sidehulls, and then we assume that the thus-augmented cushion wavemaking accounts for the sidehull wavemaking. When using this method, the key is to add an amount of cushion beam such that the added cushion lift accounts for the buoyancy of the sidehulls - in other words a beam sufficient to raise the Cushion Lift Fraction to 100%. In practice one can often get a reasonable result by simply adding the physical beam of the sidehulls to the beam of the cushion. This approximates the result described above because the sidehulls have a block coefficient less than one, but a draft greater than cushion-depression.

Of course, the accuracy of this method depends upon the sidehulls being small. If this method were entirely accurate then we could use it for the case where sidehull displacement equals the total ship weight - in other words we could use Doctors' wave drag equations to predict the resistance of a monohull. Would that this were the case!

The second method is to account for the sidehull wavemaking (residuary) drag as if the sidehulls form a catamaran in their own right. The challenge with this is that in some sense there are two catamarans involved: An inner one, representing the inner immersion of the sidehulls below the cushion, and an outer one, representing the immersion of the sidehulls below the outside free surface. We might thus wish to model two catamarans, and take their mean. And further of course, these immersions may vary with speed, making the calculation of sidehull wavemaking more tedious.

Coke-Bottling of SWATH hulls

Finally, A comment regarding the wavemaking resistance of SWATH hulls. SWATH hulls are very amenable to being modeled as a series of singularities longitudinally distributed. This in turn means that one might consider varying those singularity strengths such that the wavemaking forces cancel out, and there is no net wavemaking - at least at some critical speed. This effect is equivalent to the application of the 'area rule' to aircraft wing design, which gave rise to the Coke-bottle shape of jet fighters, and is referred to as Coke-bottling of a SWATH hull.

There is no simplified or systematic series method for estimating the resistance of a Coke-bottled SWATH hull - the only recourse is numerical methods, and I personally would start with Michlet. In general it has been observed that low speed SWATHs "prefer" low prismatic coefficient hulls (foot-long dog in a short bun), while high-speed SWATHs benefit from the use of high-prismatic hulls (dog-bone shape.) I realize that this guidance is vague! Note also that it is very different from Saunders' guidance for monohulls, given in Figure 10.1.

Conclusion regarding the wavemaking resistance of a single hull alone

In conclusion, I have presented four methods for estimating the residuary or wvaemaking component of a single hull element of an AMV. These four methods were:

- CFD

- Series data

- Series data with a Worm Curve Factor

- Model tests

In the following sections we shall estimate the other components of resistance, starting with the challenge of accounting for two or more wavemaking hulls.

9.4 Multihull Interference Drag

It would seem, to the casual mind, that the resistance of a catamaran should be simply twice the resistance of one of its hulls. Unfortunately, such is not the case. There exists an interference effect between the multiple hulls of catamaran, SES, trimaran, etc. This interference may act to either increase or decrease the drag as compared to the simple sum that seems intuitive. Interference Drag refers to an augment of drag caused by multiple hulls "talking to each other" hydrodynamically. Practically, it is found that for most multihulls the wavemaking or residuary resistance of the whole ship is slightly greater than the sum of the resistance of the several hulls measured separately.

Dr. Mustafa Insel has an excellent presentation of this in his dissertation, available online. The following paragraphs are quoted directly from his work:[32]

The resistance components of a catamaran present much more complicated phenomena than those of monohulls due to the interference effects between the hulls. The interaction effects can be divided into two distinct groups.

1) Body interference : The flow around a symmetric demihull is asymmetric due to the influence of the other demihull. i.e.the pressure field is not symmetric relative to the centreline of the demihulls. This has following outcomes:

(a) *The perturbation velocity around the demihull increases, especially on the inside, tunnel side, of the hull due to the venturi effect. This velocity augmentation causes an increase in skin friction resistance and modifies the form factor. Experiments of Miyazawa and Schimke indicate an increase in perturbation velocity of up to 10 % in the x direction compared with that of the demihull in isolation.*

(b) *A cross flow may occur under the keel which can lead into an induced drag component which is normally neglected in monohulls. Although this component is reported to be important in symmetric catamarans by Pien, Miyazawa has suggested that this component is relatively small compared with (a) in his experimental results. In these experiments the cross flow velocity in the y direction is about 5-7 % of the model speed. Crossflow in the entrance is outwards, while in the run it is inwards.*

(c) *Because the wave heights at the stern inside and outside of the demihull are different, the flow at the stern can show inwards or outwards flow. This causes vortices and spray at the stern resulting in an induced drag component.*

(d) *The velocity increase on the tunnel side may change the structure of the boundary layer.*

(e) *As the waves of one demihull reach the other hull, the wetted surface, and therefore the skin friction resistance, can change.*

2) Wave interference: As a result of two hulls running side by side, interference effects on wave resistance may also be observed.

(a) *Due to the change in the pressure field, wavemaking of the demihulls may change. In other words the wave formation of a demihull may be different than from the assumed case of the demihull in isolation.*

(b) *A favourable or unfavourable interaction between the waves of the demihulls may occur. The transverse wave of a demihull is always reinforced by the other hull while divergent bow wave of the one hull can be cancelled by the divergent stern wave of the other hull or by the reflection of the same bow wave from the other hull.*

(c) *The reflections of divergent waves from the other demihull complicate the interference phenomena.*

(d) *The bow wave of a demihull in the tunnel meets the bow waves of the other demihull on the centerline, and superposition of these two waves can become very high resulting in an unstable wave, even in breaking waves and spray at high speeds.*

(e) *Inward or outward flow at the stern changes the wave formation at the stern.*

As is clear, there is a lot going on in this hull interaction question. With a focus on practical design, the question then is "how can we deal with this? How shall we estimate or account for these factors?"

There are two solutions possible: A wave superposition technique which addresses some of the phenomena, and a more complete technique that accounts for the full interaction between hulls. Frequently - and practically - we address wave resistance of a multihull by superimposing the wave pattern of one hull - operating alone - upon the wave pattern of the other hull operating alone. This technique, however, misses all of the "body interference" effects enumerated by Insel. It is further impaired by the fact that the waves created by a demihull operating alone are different than the waves created by that same hull, when in the presence of its twin. The presence of the incident waves changes the inflow conditions upon the "target" hull, so that the waves generated by that hull are different than they would be absent such incidence.

The result of this is that the total wave system of a multihull vessel may be very different than a simple superposition of the waves generated by each hull separately.

Let us delve into the range of practical journeyman techniques for addressing this problem. Following this, I will discuss model testing techniques, and finally I will touch upon the theoretical limits of interference.

9.4.1 Methods for predicting interference drag

The simplest practical technique is to ignore interference and assume that multihull wavemaking drag (residuary resistance - I will use the terms interchangeably) can indeed be calculated by simply summing the separate hulls: We assume $Rr_{SHIP} = \Sigma Rr_{HULLS}$. We know this is wrong, but how wrong is it?

The interference drag for a multihull is well known to depend tremendously on the spacing of the hulls, and the Froude number. Some data on trimarans by Lazauskas and Tuck [5] is reproduced in Figure 9.12, and it shows that the total resistance may vary by 20% across a range of spacings. All of the ships plotted in this figure have the same length and displacement, just different configurations or positions of the ama relative to the main hull.

It is convenient to divide the interference drag into arising from two primary sources: wave pattern interference, and flow interaction. As has been discussed, this division is not accurate, but it is a usable model.

Using this model, the first of these components can be calculated by superimposing the wave trains of the several hulls, and calculating the energy of the resulting composite system.

The second of these components, the flow interaction, requires a model that is explicitly multihull. In the real situation, the waves are both reflected and refracted by the other hulls. In addition, local velocities can be affected. There is even an effect upon Frictional Resistance (see [33].)

In consequence, there are two techniques for predicting the interference drag, depending upon whether one predicts simply superposition or full interaction. For superposition one can use Michell's Integral, and the very handy 'michlet' computer program. For full interaction one must use a 3D CFD program, or model tests.

9.4.2 Model Testing Techniques

Model tests do of course completely capture the interference drag in Rr_{model} if the spacing is correctly modeled, and this is obviously the most accurate solution. But all too often, after the completion of the model tests, we decide to change the choice of building yard, and the dimensions of the new drydock force a change in beam, or weight growth forces a change in ama immersion, or the owner's requirement changes in some subtlety, such that the model-tested configuration is no longer an accurate representation of the final ship configuration. In such case it is helpful to have the ability to combine analytical methods with model tests to extend model test results to untested spacings. In this case we have to rely on some assumptions of similitude. Generally we will proceed as follows, in a technique reminiscent of the Worm Curve method:

- Use a predictive method (e.g. Michlet) to estimate the interference *factor* (not drag) of tested condition:

$$K_{int}(Fn) = \frac{R_r multihull_{michlet}}{\Sigma(R_r hulls_{michlet})} \quad (9.21)$$

- Calculate interference factor of New Configuration, at speed of interest:

$$K_{int}(Fn)_{new} \quad (9.22)$$

- Apply ratio to tested condition:

$$R_r new = R_r tested \frac{K_{new}}{K_{tested}} \quad (9.23)$$

9.4.3 Limitations

In the paragraphs above I have implied that CFD is the most accurate way to predict interference drag, and this may be true. But there is a tendency today to simply throw everything into a CFD tool and hope that the tool works correctly. I would like to caution that interference drag is one area where I have seen CFD fail to predict the drag correctly. The following discussion describes an experience encountered by Dr. Igor Mizine during the design of the HALSS trimaran, as described in his October 2009 paper "Inteference Phenomenon in Design of Trimaran Ship." [34]

Figure 110 presents residuary drag curves for two configurations of the project. The solid lines represent CFD predictions, the discrete spots represent model test results. The difference between the two configurations was only the longitudinal position of the sidehulls - in configuration 9 the sidehulls are slightly further aft than in configuration 5.

As can be seen, the CFD predicted that configuration 5 would be consistently lower in C_R over the range of speeds. The model tests do agree that it is lower, but look at the huge deviation of the model test triangles from the CFD's dashed line.

Photographs of the model tests (reproduced in Figure 9.14) give some insight into why the CFD results may be so wrong. It can be seen that in configuration 9 there is some wave breaking taking place in the main hull's stern wake, that is not captured in the CFD. How were we to know this would happen? What if we had relied on the CFD and not conducted the model test? At 35 knots there is a 50% error in the C_R.

9.4.4 Theoretical Interference Limits

Tuck & Lazauskas [5] have written a very interesting paper exploring the theoretical limits of interference drag for multihulls. Their work is available online from cyberiad.net. The work contains several innovations that are worthy of discussion. The first of these is the invention of a single parameter - σ - which can be used to map a space including monohulls, catamarans, and all possible trimarans.

Lazauskas' 'σ' is defined as 'the ratio of the displacement of all the outriggers divided by the displacement of the total ship.' Thus, in the case of a monohull, the displacement of the outriggers is zero, and σ is zero. In the case of trimaran, where the main carries 80% of the ship weight, and the outriggers each carry 10%, the value of σ will be 0.2. And in the case of a catamaran the outriggers carry all of the weight - there is no center hull - and thus σ is 1.0.

Lazauskas and Tuck used Michell's Integral to estimate the drag of multihulls of a standard one cubic meter displacement. (These may then be Froude scaled to any desired size.) They studied vessels of all σ values from 0 to 1. Their 1996 paper [5] presents many interesting results, a few of which are reproduced here.

In Figure 9.15 we see the results of total drag (C_T) for all σ, for three Froude numbers. A few results leap out, which I feel are fitting observations to end on:

- At all speeds the unconstrained monohull ($\sigma=0$) is superior to any of the multihulls.

- The catamaran ($\sigma = 1$) is better than any trimaran having $\sigma > 0.2$

Worded in the imperative tense:

- If you can, design a monohull

- If the monohull is 'out', then design a catamaran

- If you must design a trimaran, keep the outriggers small, say below 10% of the displacement each

This result by Tuck and Lazauskas is very attractive for it's simplicity and it's clarity. We need however to recall that their work was performed using Michlet, which is a wave superposition technique and therefore does not fully capture the hydrodynamic interaction between the hulls. Mizine, in his October 2009 paper "Interference Phenomenon in Design of Trimaran Ship" [34] demonstrates the feasibility of exploiting the interactions for a practical trimaran with deep SWATH-like amas. For his ship, as the result of the application of sophisticated modeling capabilities (outside the range expected of an undergraduate student) he is able to virtually eliminate the drag of the sidehulls. His conclusion states *"...a trimaran can be designed such that favorable hydrodynamic interactions offset almost all of the side hull drag over a practical range of speeds. For example at the 32 to 34 knot speed the resistance of the trimaran (Experiment 5) is equal to the resistance of the center hull (Experiment 3.) In the trimaran configuration there was 18% more displacement yet the drag was the same as that of the center hull."*

Mizine's 18% increase in displacement is equivalent to a σ-value of $\frac{0.18}{1.18} = 0.15$. By looking at Figure 9.15 we see that Tuck & Lazauskas' results would lead us to expect a C_T increase of perhaps 20% for this σ-value, compared to the value for the $\sigma = 0$ "center hull alone." Mizine's result is that the drag is the same in these two cases, despite the increase in displacement and wetted surface - in other words C_T went *down* not *up*.

This result reinforces our comment that superposition methods are a good way to start, and are a good journeyman technique to be used, but that they are not a complete solution of the problem - and by Mizine's own experience only the physical model test is reliable.

9.5 Lift System Air Momentum Drag

For the powered lift craft there is a volume of air that is taken from the atmosphere and stuffed into the cushion. (It then leaks out of the cushion and is replaced with more air.) But in the process of being taken from the atmosphere (which is nominally at rest) and placed into the cushion (which is moving with the ship) it must be accelerated from rest to ship speed. This requires power, and this power is expressed as a drag due to lift air momentum. Of course, it only applies to air cushion vehicles.

Note that the act of accelerating the air from rest to vehicle speed is done at the air inlet. So if the inlet is not located on the ship, then this drag is also not present on the ship. This consideration only applies to models, but it is an important one in those cases where, due to the size and weight constraints of the model, the lift fans are mounted on the tank carriage, and the lift is then ducted to the model using "dryer hose." In this case the model drag will not include the lift air momentum drag, and will thus tend to underpredict the ship drag.

The lift air momentum drag is an inertial problem, so it Froude scales or "λ-cubes" from model to ship. However it depends upon the lift air flow rate, and the lift air flow rate at the model may be different from that at the ship (even apart from scale considerations.) This may be because the model scale fans have different characteristics, or because it is hard to set the model stern seal to the perfect inflation condition, or due to differences in myriad other parameters.

I therefore feel that the lift air momentum drag should be scaled independently: It should be subtracted from the model drag (by calculating it based on model flow rates) and then re-added to the ship drag after calculating it based on ship flow rates. Of course, in the simplest case we will assume that the ship flow rate is λ^3 times the model rate, so the effect is nil. But it is a good habit to get into even in this simple case, because of the variations in flow that will undoubtedly enter as the project continues.

Lift air momentum drag is fully predictable if you know the lift air flow. The drag is correctly handled by calculation, as follows:

$$R_{momentum} = M_{air}\delta V_{air} = \rho_{air}QV \qquad (9.24)$$

Where:
Q = Lift air flow (m^3/sec)
V = Craft velocity (m/s)

9.6 Skirt Drag

Air cushion vehicles, such as SES or ACV, also have fabric skirts in contact with the water, and these cause drag. The drag of skirts over water is not well understood. Recent work by Larry Doctors [35] (which will be described below) is making breakthrough understanding of this drag component, which is turning out to be more complex than previously understood. As a result I am inclined to say 'watch this space' for further developments in prediction of this item.

The earliest and simplest method for modeling skirt drag was to assume that that drag of the skirt was simply friction: Some amount of fabric is being dragged across the surface of the water. The calculation that ensues is then fairly simple: Estimate a wetted length and calculate the corresponding Reynolds number. Determine a C_f for this Reynolds number. Estimate the wetted surface, and use the C_f and the wetted surface to determine a skirt drag.

Now, because as we well know the frictional component can't be scaled directly from model test, what we have to do is perform the above estimation at model scale, determine a predicted model-scale skirt drag, and subtract that from the measured model drag before proceeding with the rest of the model test extrapolation. We then determine an estimated full-scale skirt drag which we re-add to the estimate during the recomposition phase.

The problem is that skirt drag is not purely frictional - there is at least some component that might be considered 'residuary.' If this part accounted for 100% of the skirt drag, then we could simply leave the skirt drag alone, lumping it in with other residuary components of the model test.

In my own early work [24] I suggested using both methods and thus "bounding the problem." In other words I would extrapolate a model test both ways (skirt-as-friction and skirt-as-residuary) and would then get two different full-scale drag lines. I would then believe that the truth lay somewhere between them.

Of course, the above method works for model test extrapolation, but what about for drag estimates performed during early stages? Certainly the skirt-as-friction model can be used to estimate a skirt drag, and this is better than nothing.

In recent years Doctors [35] has developed a new model of skirt resistance that is much more comprehensive - and is also providing results that track very well against model test measurements. His most recent (2011 - unpublished as yet) analysis of model test data suggests that the parabolic deflection model gives the most accurate results.

Doctors' solution is described as follows. First, consider the geometry of the bow and stern skirts of an SES. Doctors proposes the generalized geometries depicted in Figure 9.16.

The stern seal is modeled as simple friction, as discussed above. Doctors provides equations for solving for the contact length based on static considerations of the pressures. He then modifies the geometry by adding the effect of the stern seal drag - the drag force on the bottom of the bag will pull the bag slightly aft, increasing the front radius and changing the balance of forces.

This is a good model, but it is difficult to capture one of the other realities of stern seals: When ideally tuned, they don't in fact touch the sea surface. They glide with a tiny daylight gap visible above the water - and hence have no friction.

The bow seal is much different. By Doctors' new theory the drag due to the bow seal has two components, a viscous (frictional) component and a wave Pile-Up component. This is equivalent to saying that $Rt_{skirt} = Rf_{skirt} + Rr_{skirt}$. Where, of course, Rr_{skirt} will properly scale from model tests, but Rf_{skirt} must be subtracted and extrapolated independently.

This is a new theory and is not embedded in current model test standard procedures (such as the ITTC guidelines.) Current standard procedure is not to separately identify Skirt Drag. A half-way measure would be to estimate a skirt wetted area and add it to the hull frictional extrapolation. But a journeyman practitioner who finds himself supervising an SES model test program would do well to impose some special procedures for extrapolating skirt drag.

9.7 Air Cushion Wavemaking

In the quest for speed we introduced air cushion sustention. The air cushion replaces the rigid hull of the ship with a bubble of air, and this bubble has demonstrably no friction. But this bubble does still push the water out of the way as it passes, and in so doing it generates waves. Remember, it all boils down to $F = MA$. The air bubble pushes some water out of the way. To do this it must accelerate the little water particles. They have mass, so this takes force. The result is a drag force, the manifestation of which is waves.

Hydrodynamically, the bubble of air replaces a "known geometry / unknown pressure" boundary condition (a typical ship hull) with one where the pressure is known (equal to the bubble pressure) but the geometry is unknown. It can be interesting to study the wave patterns generated by air cushions, but this is beyond the undergraduate journeyman level. At the practical level, what we care about is a means for determining the wavemaking resistance due to the air cushion.

Newman and Poole [36] were the first to solve the mathematics of this for practical use, and they produced curves of cushion wavemaking drag versus speed, as reproduced in Figure 9.19.

Several features are noteworthy. First, let's familiarize ourselves with the axes used: The speed axis is Froude number, but this is a Froude number based on the square root of the cushion area (area = length x beam.)

The force parameter is Drag over Lift - the 'L' in the numerator of the y-axis is the cushion lift, which is equal to pressure times area. In the denominator is a term of Pressure over square root of Area. This of course becomes a density - pounds per cubic foot, say - and is referred to as the cushion density. Given all other things being equal, cushion wavemaking drag increases linearly with cushion density.

Now let's look at the data curves themselves. There are seven full curves plotted, corresponding to different cushion length-to-beam ratios from 2 to 8. At low speed, say $Fn = 0.8$, the trend is as we might expect - a slender cushion $(L/B = 8)$ has a lower drag. But at high speeds, above $Fn = 2.0$, it is the short fat cushion that has the lower wavemaking drag parameter.

Note that the Newman and Poole data is cut off at a speed of about $Fn = 0.6$. This is because below that speed the mathematics shows great instability. Figure 9.20 presents an enlargement of the very low speed region. (Note the x-axis in this figure, which is an inverse Froude number such that high speed occurs at x=0.) Doctors seminal work in the early 1970s [37] was to introduce smoothing parameters into the solution, recognizing that the cushion pressure can't instantaneously rise from zero to full, but there must be some 'ramp up' of pressure with distance. This is depicted in Figure 9.21.

As a result of the smoothing parameters Doctors produced a new set of equations for predicting the wave drag of the cushion. Numerical results of these, similar to the graph in Figure 9.19, are presented in the graph in Figure 9.22. Note importantly that the Froude Number in the graphs is based on the cushion area (its square root, actually), so any craft of a given cushion area and speed would have the same FN, despite having differing L/B ratios.

Let us consider again the importance of the L/B sensitivity in these results. As mentioned earlier, for high speeds the data shows that low L/B cushions have lower wavemaking resistance. At low speed the higher L/B forms are indicated.

Figure 9.23 reproduces a Navy study of four different L/B choices for an 8,000 ton SES, from Reference [38]. The curves are of total resistance, not cushion wavemaking only, but this does not change the picture. It is clear that from the point of view of resistance, in speed range 'A' the long slender cushions are desirable, with L/B of 6 or 8 being nearly equal in performance, while at very high speeds - speed range 'B' - the more box-like L/B=2 form is greatly superior.

Finally, note that the drag in Doctor's graph is non-dimensionalized on Pressure-Squared. Because of this Pc^2 effect, the L/B for minimum R_W is rarely the same as the L/B for minimum C_W. That is to say, that not only does L/B vary in a given ship optimization, but usually the pressure does too. Even if two configurations have the same area they may not have the same weight, and thus they may not have the same cushion pressure. So then to choose between them we glance at the Doctors curve and see which configuration has the lower Cw. But ah ha! Maybe the lower Cw is the ship with a higher pressure, and maybe the difference in pressure-squared is bigger than the difference in Cw! In this case the ship with higher Cw might have lower total Rw. The point of this is that optimization studies must be carried forward all the way to dimensional drags and powers, and not be performed at the level of non-dimensional coefficients.

188

9.8 Spray and Spray Rail Drag

High speed craft may generate significant amounts of spray. According to Faltinsen this can be 12% of the total resistance of the craft. The interested student is directed to Faltinsen [4] page 36.

Spray drag consists of two components, a Pressure Drag a function of Froude number, and a Frictional Drag a function of both the Reynolds number (Rn) and the Weber number ($Wn = \rho V_{spray}^2 d_{sr}/T_s$, where $d =$ spray thickness, $T =$ surface tension.) The problem lies, once again, in our inability to scale such parameters as surface tension of the water. If we try to apply Reynolds scaling, we don't know what velocity the flow has - it is certainly not the same as the ship speed! If we try some development based on the Weber number, we don't know the thickness of the spray sheet. Even earlier in the design process, there are no good and simple predictive methods.

Our solution is to try to avoid this entire challenge. To return to my Lewis & Clark metaphor: When faced with a river we can't cross, because we can't build a bridge out of the knowledge that we have, we will have to try to find a route that takes us around to where there is no river. In our case, this means that we try to minimize spray generation by using spray rails on the hull.

The best guidance I have seen on practical design of spray rails is given in Faltinsen, derived from work by Muller-Graf in 1994. His guidance boils down to the following:

- Spray rails start 3% of LWL above LWL at FP

- Taper to LWL at midships

- Spray rail width about 0.6% LWL for slender hulls

9.9 Appendage drag

Advanced marine vehicles may have appendages, and these appendages do have drag. There is nothing particularly AMV-unique about these, so this section is quite brief: Treat AMV appendages by using the same tools as are used for appendage resistance estimates on conventional ships.

The key point to be made here is that the model testing of AMV appendages may be even more unreliable than the already-difficult challenge of the conventional ship. This is because the high speeds of the AMV mean that force generators such as rudders can be made quite small. This smallness exacerbates the scaling challenges that are already well known (as well as exacerbating the difficulty of the model making, itself.)

I recommend that it is better to have a bare model and handle appendages by calculation. The uncertainties introduced by this method are not likely to be any greater than the uncertainties inherent in scaling tiny high speed appendages.

9.10 Rules of Thumb

This might belong in the earlier discussion of performance metrics, but it is not an overall metric, merely an estimator of catamaran resistance and propulsion.

In the April/May 2011 issue of Professional Boatbuilder [39] Mr. Derek Kelsall (FRINA, Kelsall Catamarans Ltd, Waihi New Zealand) contributed a very simple formula for predicting the performance of powered catamarans. According to Kelsall this formula works quite well for "typical cruising multis."

$$Vk = \sqrt{\frac{HPLWL}{\Delta}} \tag{9.25}$$

Where:
Vk = Speed in knots
HP = Horsepower
LWL = Length in meters
Δ = Displacement in tonnes
Expressed another way, this is equivalent to:

$$\Delta \frac{Vk^2}{LWL} = HP \tag{9.26}$$

In other words, displacement times the square of the Speed-to-Length ratio yields Power. This might be a very handy early-stage estimator. Note, however, that it does not contain any guidance on the selection of length for a given displacement and speed. Attempts to combine this with McKesson's "Observed Best Attainable" relationship between speed and displacement, and thus generate a target value for length, yield interesting results. The reader is encouraged to try it herself! The useful part of the result is that, for speeds in the range $1.0 < FN_L < 2.0$, the length in meters should be 75% of the speed in knots.

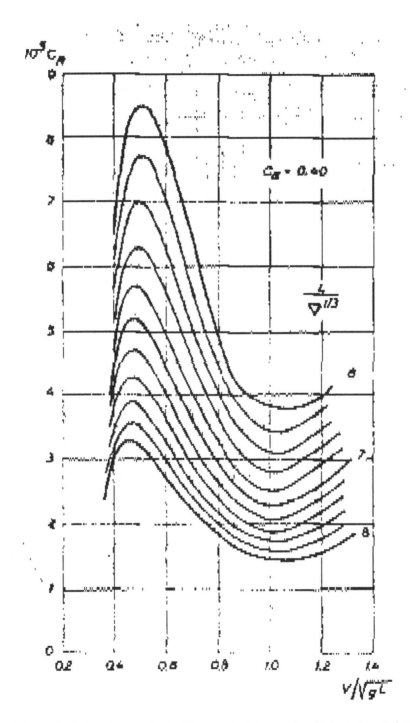

Figure 9.11: Contours of Residuary Resistance Coefficient for B/T $=3$
$CB = 0.40$ from the Lundgren series [6]

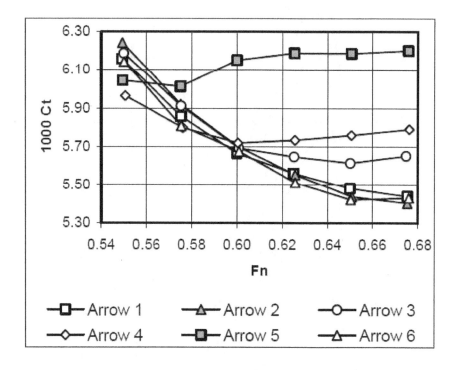

Figure 9.12: Total Resistance Coefficient for six Arrow Trimaran configurations, from Lazauskas and Tuck [5]

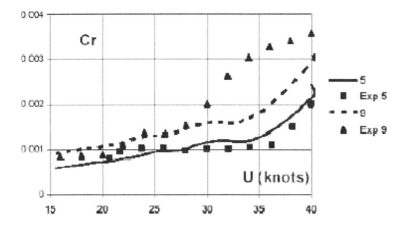

Figure 9.13: CFD and model test results, for a 2009 study of the effect of longitudinal position of side hulls on trimaran residuary resistance

Figure 9.14: Comparison of the free surface behind trimaran 5651 in Experiment 5 (left) and Experiment 9 (right) at Froude Number = 0.34

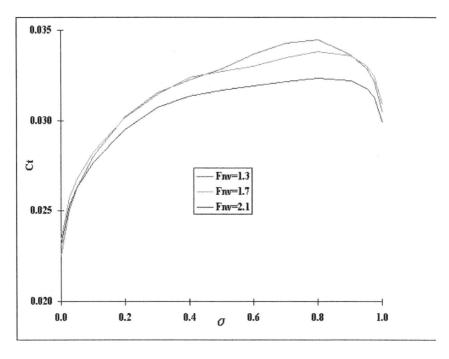

Figure 9.15: Total resistance of optimized one-tonne generalized trimarans
[5]

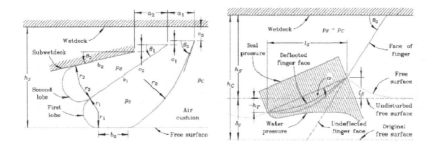

Figure 9.16: Doctors' geometry definition sketches for a stern seal (left) and
a bow seal (right)

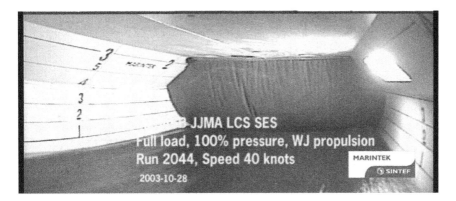

Figure 9.17: An SES stern seal exactly corresponding to Doctors' definition sketch

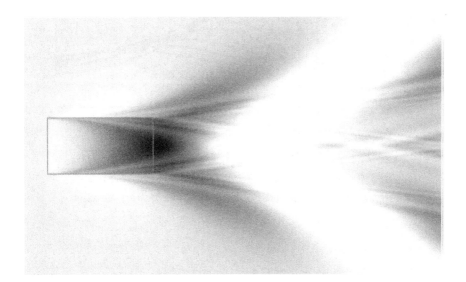

Figure 9.18: The wave pattern caused by a rectangular constant-pressure patch

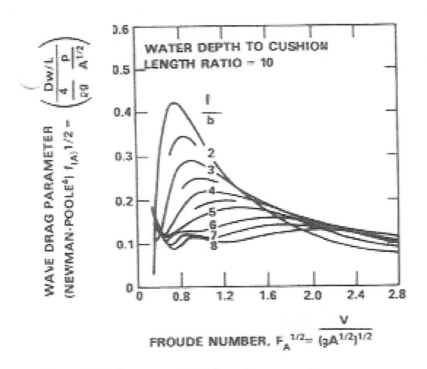

Figure 9.19: Newman and Poole cushion wave drag parameter

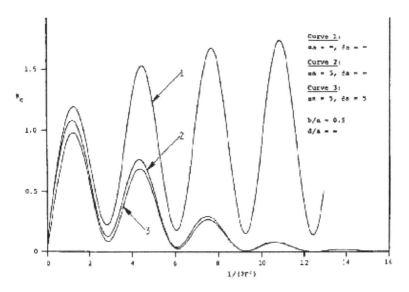

Figure 9.20: Doctors' figure showing the Newman and Poole instability, and the smoothing accomplished by introducing parameters alpha and beta

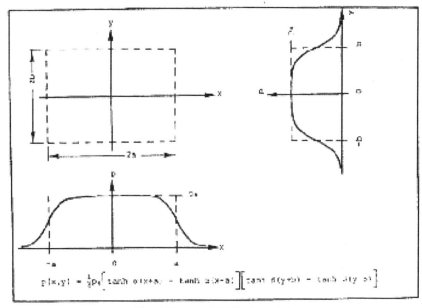

Fig. 1 Pressure distribution used

Figure 9.21: Doctors' pressure smoothing parameters

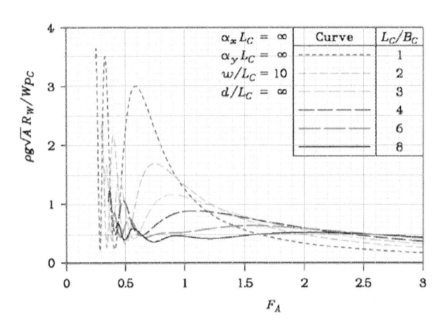

Figure 9.22: Doctors' results for cushion wavemaking drag

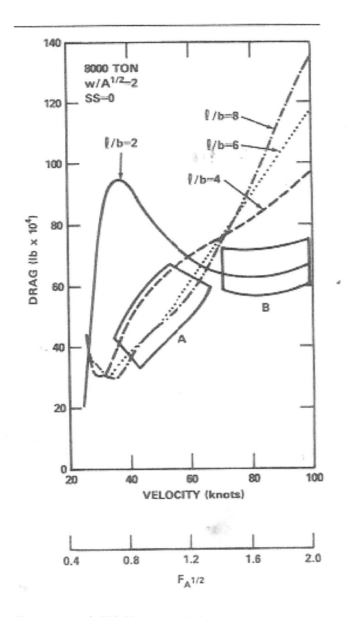

Figure 9.23: A US Navy result for total drag of an 8,000 ton SES as a Function of Speed and L/B ratio [7]

10 SWBS 070 - Hull Form Design

How do we pick the hull form parameters for the following types of vessels?

- Catamaran

- Trimaran

- SES

- SWATH

In each case, I will use a purpose-driven approach to hull form development. Once we discuss the purpose of each of the hull form elements, then we can seek parent forms for those hulls, and then we can develop a design procedure.

10.1 Catamaran hulls

10.1.1 Catamaran hull form teleology

What is the purpose of a catamaran hull? For all buoyantly-supported craft the first requirement is that the hulls displace a volume of water equal to the craft's weight. But beyond this, let's return to the purpose of the catamaran: A catamaran is a way of getting extreme hull slenderness while still having acceptable stability. And the extreme slenderness was sought in order to reduce resistance. So the primary purpose of a catamaran hull is to have low drag. But there are very important secondary purposes that must be worked in as well. The hull form must minimize the occurrence of slamming on the cross structure. The hull must also be wide enough to fit the propulsion machinery. I submit that these three are the top-level catamaran hull form teleology:

- Minimum Drag

- Minimize Slamming

- Fit the Machinery

10.1.2 Catamaran hull form parents

There are few published hull form series intended for use as catamarans. The only publicly available Catamaran systematic series I know of is the German VWS 89 [29]. Usually, designers collect their own parent data, especially collecting and systematizing the data from each catamaran of their own design.

Fortunately however all of the monohull systematic series can be used with proper accounting for interference effects. The same is true of the monohull extrapolation / offspring techniques, again as long as there is proper accounting for interference effects. This then opens up a wide field, wherein we can write down a catamaran hull form development procedure that uses a standard naval architectural data base.

10.1.3 Catamaran hull form development procedure

The reader can already see, from the foregoing, that catamaran hull form design procedure is going to follow the same rules as displacement monohull design. The one early deviation is that machinery size will probably define the hull beam.

If water jet driven, then the waterjet mounting diameter will establish the transom beam. The main engine width (and spacing, if multiple engines per hull) will define the beam slightly further forward. Once these beams are established, and of course the required displaced volume is known, then the design proceeds: The Sectional Area Curve is your key design tool. Traditional targets for Prismatic Coefficient and Fatness Ratio are very useful (see Saunders' guidance [40], reproduced as Figure 10.1)

The design of immersed transoms is an interesting area that is not well treated in mainstream literature. My personal techniques are derived from reviewing texts from the 1940s on the design of high speed displacement motorboats. The gist of the method is this:

Develop a sectional area curve for a fictional hull that operates at your target speed, but is longer than your ship. Set the parameters of this sectional area curve in accordance with Saunders" guidance, etc. Then simply truncate the curve at the desired length of your ship. Use the resulting forward portion as your ship's S.A. curve.

As a final check, design your ship's transom such that it has a draft that ensures the transom is dry at the design speed. As a transom drying criterion I use the requirement that the Froude number based on transom draft must be 2.5 or greater at design speed. (Froude number on transom draft is simply $FN_t = \frac{V}{\sqrt{gT_{20}}}$ where T_{20} is the transom draft).

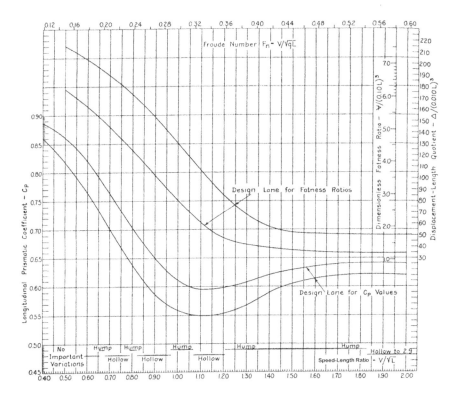

Figure 10.1: Saunders' guidance for the selection of desired Cp and Fatness Ratio

10.2 Trimaran Amas

The main hull of the trimaran may be designed by the same procedure as the hulls of a catamaran. But what of the amas, or outriggers? Let us follow our same pathway through development of the form of these hulls.

10.2.1 Trimaran Ama hull form teleology

Pause and consider: What is the purpose of the trimaran? What is the guiding concept? A trimaran is a very slender ship, so slender that she would be unstable unless sidehulls (amas) were added. So the purpose of a trimaran ama is stability, and very little else. If we didn't need the stability of the amas, we wouldn't have amas at all.

So next, let's remember our Sophomore-year lectures on stability: Stability is all about waterplane area and waterplane inertia.

From this argument we see that the real purpose of an ama is to add waterplane inertia to the ship, by adding waterplane area. Displacement - *per se* - is not needed in the ama. At the same time, we want to minimize the drag of the ama. What is the hull form with minimum drag? Clearly it is a hull of minimum volume and minimum wetted surface. Combining these thoughts, we see that an optimal ama would:

- *Have* waterplane area

- *Not have* displacement

- *Not have* wetted surface

Obviously the first and last of these are in conflict - the minimum shape that satisfies this target would be a flat plate, having no draft, and having the minimum possible wetted surface for the given amount of waterplane area. Now, in a real ship we also need the amas to perform across some range of loading conditions, and some range of ship motions, so we do need them to have draft. But clearly we can see the trend: We want amas that are shallow, amas that have high B/T ratios.

10.2.2 Trimaran Ama hull form parents

The argument above, based on teleology, is my own. I should hasten to state that there is no consensus on Ama form: I tend to prefer moderate L/B round-bilge semi-planing forms, Dr. Tony Armstrong (Austal) - who has extremely good credentials in this area - prefers very slender high L/B forms. Dr. Igor Mizine (CSC) - who is also a recognized expert - prefers SWATH forms.

I will begin with my own logic, and will then attempt to do justice to these other points of view.

From my teleological argument the suitable parent forms for trimaran amas are shallow hulls maximizing the amount of waterplane area for each ton of displaced volume. The amas will also be kept as small as possible, which means that they will operate at a higher Froude number than the main hull. Finally, remember Lazauskas' results on "optimum multihulls" [5] which suggest that the amas should not be more than 10% of the total ship displacement (e.g $\sigma < 0.2$).

This leads to the selection of round bilged planing hulls as parents for the amas.

I like to use Series 64 for this purpose. Series 64 has a nice high B/T value, operates at the right Froude numbers, and is widely available in standard naval architecture software and reference materials [28].

Dr. Armstrong attains the same goal, but rather than using a planing-like form he continues the trend of slenderness and uses a sharp-veed long narrow form - see Figure 10.2. This is apparently because of his experience with the need to accommodate a range of drafts at which the ama provides its waterplane area (a range of drafts is needed because of the loading and damaged cases for the ship stability. My argument of teleology may be claimed to be simplistic, because it treats the ship as if stability is only needed at the design condition.)

Dr. Mizine takes a very different approach - see Figure 10.3. His amas are narrow and deep, and may include SWATH-like bulges at the bottom. Part of his motivation is because he likes to fit machinery into the amas, and the SWATH-like form provides very good inflow to a submerged propeller, providing good propeller efficiency.

He has further found that by carefully positioning the amas, depending upon their volume and the ship speed, he can cause favorable interference effects that completely offset their drag: That is to say that the resistance of the whole ship is no greater than the resistance of the center hull alone - the amas are "free."

Unfortunately the tools needed for this optimization are beyond the undergraduate level of this present work, but the concept is very interesting and is being increasingly documented in Dr. Mizine's growing body of published works. (For many of these, search the website of the Center for Commercial Deployment of Transportation Technology, http://www.ccdott.org/ www.ccdott.org)

10.2.3 Trimaran Ama hull form development procedure

Based on McKesson's philosophy of ama design, the following procedure obtains:

Given main hull:

- Estimate KG

- KG yields GM-required

- GM yields BM-required

- BM yields I_T-required

- I_T-required defines a 2D relationship between:

- Spacing (d)

- Waterplane Area (A_{wp})

- For each selected Waterplane Area, now must decide what L & B to attain it

- Have a realistic ama-draft (T_{ama}

- What combination of L, B, T has the desired A_{wp} and minimum drag? See Saunders' design lanes and other traditional tools

10.3 SES Sidehulls

Now we turn to a different type of craft. The Catamaran and Trimaran are both Buoyant Lift craft, and their hull form development is a lot like the design of conventional ship hulls. In the case of the SES this is no longer true - the presence of the powered-lift cushion dominates the design of the hulls.

10.3.1 SES Sidehull hull form teleology

The primary purpose of the SES sidehull is to retain the cushion. The sidehull must extend down below the bottom of the bubble - at all speeds, cushion pressures, craft attitudes, sea conditions, etc. This translates to a draft requirement.

We also fitted sidehulls to the SES (as opposed to being a fully-skirted ACV) because we wanted to fit marine propulsion. Therefore the sidehulls need to accommodate the propelling machinery of the ship. This translates to a beam requirement (or a set of beam requirements, at various locations.)

The sidehulls' shape is constrained by their need to avoid interference with the fabric skirt systems of the craft. These skirts will be discussed in a later unit of this course, but the point to take here is that the skirts require that the sidehulls be completely wall-sided vertically and completely straight longitudinally in the region of the skirts. Transitioning into and out of these straight-line sections can be a challenge if a radical sidehull shape is chosen. This yields the "standard" SES sidehull shape which is wall sided and straight-lined on the cushion side over it's entire length.

When off cushion (normally at zero speed, but sometimes off-cushion operation is conducted with some small ahead speed) the SES becomes a catamaran. It must in that case float on its sidehulls, so they must have a deep-draft volume equal to the ship's weight. Further, their LCB in the off-cushion condition must be aligned with the ship's LCG at some acceptable trim. (Normally, SES off cushion float with a substantial trim by the bow. But even attaining this degree of trim requires attention to the location of these centers.)

Figure 10.2: Gives some depiction of the form of Ama preferred by Dr. Tony Armstrong

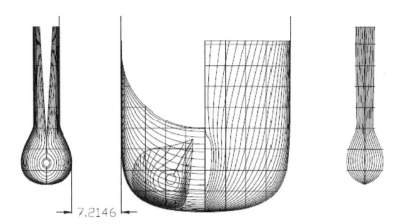

7.2146

Figure 10.3: A depiction of the SWATH-like Amas preferred by Dr. Igor Mizine

There is a different set of centers when the craft is on-cushion. In the on-cushion mode the sidehull LCB is that corresponding to a much lower draft, but there is also a longitudinal center of pressure - designated LCP - which represents the effective location of the cushion lift. In the on-cushion case the craft trim will be the result of the confluence of these three centers: LCG versus LCB+LCP. The design of the sidehull needs to accommodate this.

Finally, the sidehulls of the SES are the sole source of the ship's transverse stability. This too is the subject of a later unit, but we will foreshadow it here by saying that there are two components to this stability: Static stability of the conventional metacentric type, and dynamic stability due to planing forces on the hull bottom, in high speed turns.

And of course, we want the sidehull to perform all of the above tasks with minimum drag - thus minimum wetted surface.

SES sidehull teleology may be summarized as:

- Retain the Cushion (e.g. Draft)

- Accommodate the skirts

- Fit the machinery

- LCB / LCP / LCG alignment

- Metacentric Stability

- Planing Stability

- Minimize Wetted Surface

10.3.2 SES Sidehull hull form parents

There are arguably two classes of SES parent hull, although only one of them is seen today. The rare one is the so-called "lenticular" hull. This is a hull having curved waterlines on the outside, resulting in a hull almost identical to that of an early HobieCat pleasure boat. Lenticular hulls were developed in the 1980s as solutions for SES having relatively modest speeds (e.g. thousand-foot craft of 50 knots). There are no lenticular-hulled SES that I know of afloat today.

The more traditional hull is "prismatic." This means that it has a simple geometric shape that is continued over nearly the entire length of the hull (except for a transition at the bow.) The key hull form parameters for the prismatic hull are Sidehull beam, Sidehull draft, and Deadrise angle.

10.3.3 SES Sidehull hull form development procedure

The simplified statement of the design procedure for a prismatic SES hull is:

- Find Draft to retain cushion, including wave effects

- Find maximum acceptable deadrise angle for planing stability

- Find Beam to yield desired metacentric stability

- Include machinery haunch if needed

- Include waterjet or propeller fairing if needed

How do you decide the sidehull dimensions? Let's start with sidehull draft: Retaining the cushion is of course Job No. 1 for the sidehull. And that means Draft, both Outer and Inner. Outer Draft is the draft from the ocean free surface to the keel of the sidehull. Inner Draft refers to the wetted draft below the bubble of the air cushion. Outer Draft is equal to the sum of the bubble depression and the Inner Draft.

For a starting point, you want an inner draft that is 30-50% of the bubble depression - that is to say that at 1 meter of cushion pressure, the sidewall would be 1.3 - 1.5m deep, so as to have that 0.3-0.5m "fence" for the cushion.

This factor of 30-50% is of course simply a rule of thumb with no physical basis. The real physical basis would be to model the wave shape of the cushion-generated wave. You need the sidewall deep enough so that the trough of that wave doesn't vent. That trough is deepest right at hump - it's where the hump comes from, physically.

But if it does vent, you can simply dial down the cushion pressure, reducing the bubble depression and settling the sidehulls a little lower. In fact, this has the effect of reducing cushion drag so dramatically that most SES do this as a means of easing their way through the hump regime, rather than simply "blasting through"' on power.

So then the next limit is to think about what ocean waves will do to the bubble depression, at high speeds. At Fn=infinity there is no ship-generated wave. And indeed if you look at videos of SES cushions, there's really not much cushion-wave at about 40 knots - the bubble is flat. So now the question is how much ocean waves will change that, and I don't know the answer to this. In practice, add a few inches of draft to account for this effect.

So that defines draft. Then comes sidehull beam. Two concerns are tempting: (1) keeping sidehull displacement at a target value and (2) classical naval architecture issues, like L/B ratios, slenderness, waterline entrance angles, etc.

Regarding (1) I think frankly that it's a red herring. After all, an SES (probably) has less resistance the higher the cushion fraction - we'd go to 100% if we could do that without losing the air everywhere. (ACVs have less drag than SESs, but they take too much lift power because they vent all the way around.) So I think that really what you want is sidehulls with minimum drag, and this will mean minimum wetted surface. In the limit, if draft is fixed, then the minimum wetted surface is a flat plate that sticks vertically down to that draft. Any amount of deadrise or thickness will simply increase the girth, and hence increase the wetted surface, for the fixed draft.

Now, a flat plate sidehull would also be nice for LCB / LCP / LCG alignment, because the LCB would be amidships, right where the LCP is. The more shape the sidehulls have, the more they are going to be triangular, resulting in aft-shift of the LCB.

Of course, a flat plate is hard to fit the machinery into, and we will see that it has dangerous stability implications too.

So in reality we have some sidehull thickness or beam, generally based on some deadrise angle. As we will discuss under the heading of SWBS 079 - Stability, the deadrise angle needs to support planing stability: You want the vector, normal to the deadrise surface, to pass above the VCG of the ship. This sets an upper limit on deadrise.

Now, that last point becomes critical, because it means that sidehull beam will need to change if cushion beam changes. Consider:

As the cushion beam comes down, draw a midship section, and draw a vector from the keel to the VCG at centerline. As the beam comes down, this vector gets more and more vertical.

That vector represents the hydrodynamic lift on the planing surface of the sidehull. If it passes BELOW the VCG the boat trips and rolls over in a high speed turn (flip ahead to Chapter 13 to see this illustrated.)

Now, for minimum wetted surface, you want the planing surface to be a straight line that extends from the keel to the waterline. But if you draw that at some angle, say 45°, then at some beam the 45° deadrise will send that vector too low, and you'd need a lower deadrise, say 30° or something. For this deadrise to reach from the keel to the waterline, the only possibility is that the sidehull beam has to be greater.

Thus as cushion-beam comes down, the vector needs to be more vertical, thus the deadrise needs to be lower, thus the beam needs to be greater. But greater beam and lower deadrise will yield higher wetted surface, and will manifest themselves as excessive sidehull buoyancy.

So in this unit we need to also consider how to pick the cushion beam. We have already seen that the cushion Length-to-Beam ratio will affect cushion drag, so presumably we have picked a target value for L/B ratio. But this value can be respected at an infinite number of lengths and beams. Let us turn our attention to the old ACV parameter called "Pc/L" (pronounced "P C upon L"). If you imagine a profile of the bubble, you'll see that this is kind of a "draft to length ratio."

Now, imagine a tanker hull: The profile is practically rectangular, with a vertical stem leading to a radiused forefoot. But the good naval architect paid a lot of attention to his waterplane shape, and maintained his waterline entrance angle to a nice fine point, say 5 - 10 degrees.

An SES on the other hand, has a nearly rectangular waterplane, but we'd like to put a "point" on our profile - it's the tanker turned on its side. And the only way to put a point on the bubble is via Pc/L. (Ignoring ideas like segmented cushions!).

As a benchmark consider a limit of Pc/L = 100 Pa / m . This results in a Length-to-Draft ratio of 100:1. So for a ship with a cushion length like 70m, this would suggest a cushion pressure like 7kPa, or 0.7m bubble depression.

One can certainly go higher than that. How much higher? So what? The effect, after all, is only drag, just like putting too blunt a bow on that tanker. What's the waterline entrance angle of a 1000-foot Laker? Darn near 90° it looks like! Sometimes you just bite the bullet and do what you have to. But I'm sure the laker designers would like to have pointy-er bows, just like we'd like to hold our Pc/L down to 100 Pa/m.

In 1975 Mantle [41] suggested limits of 12.5 to 20 ft/ton^(1/3), which works out to 43 (low density craft) to 177 (high density craft) N/m^3 (or Pa/m - they're the same unit.)

Of course, picking length and pressure drives us (for a given weight) to a choice of beam, and we then compare that to the needs of the rest of the design in a normal naval architecture tradeoff. There is most definitely *not* one single solution to a given set of requirements.

10.4 SWATH Hulls

SWATH ships are designed to minimize ship motions. The purpose of a SWATH is to provide a ship that is decoupled from the waves on the sea surface. We do this, conceptually, by putting the ship's buoyancy well below those waves, and then supporting the human-occupied part well above the waves, attached to the submerged buoyancy by struts.

That description works as well for a semi-submerged drill platform as it does for a SWATH. And indeed, a semisub of that type is in fact a SWATH, but one optimized for zero speed.

In this course we will concern ourselves with SWATHs optimized for some non-zero speed, which introduces the need for hydrodynamic shaping of the hulls and the surface-piercing struts. This takes two forms primarily: That of selecting the optimum prismatic coefficient for the speed of interest, and then the more advanced method of "coke bottling" the hulls to minimize drag at one speed.

In order to understand the design of SWATH hulls for minimum motions, we need to have a couple of additional terms in our lexicon: "Platforming" and "Contouring." Contouring motion is the motion when the ship follows the contour of the sea surface - up and down along the waves, more or less maintaining a constant height to the water, like a cork. Platforming, on the other hand, corresponds to the ship maintaining its height relative to the earth, and letting the waves pass beneath it without responding to them - it is like a "platform" on the seabed.

A SWATH will operate in both of these modes naturally. If we think of a SWATH as a simple spring-mass system, we can imagine that at a very low excitation frequency the ship will simply move in 1:1 correspondence with the excitation - contouring. On the other hand, at a very high excitation frequency the heave mass of the ship "can't respond fast enough" and virtually ignores the excitation - platforming. As this explanation shows, the transition between these modes is governed by the relationship of excitation frequency and natural frequency, called "Tuning factor." This relationship is depicted in Figure 10.4.

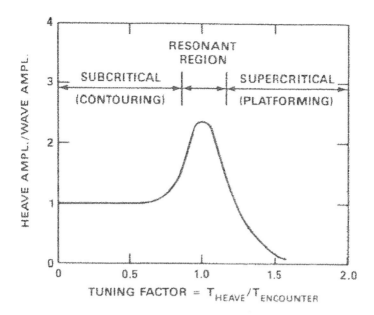

Figure 10.4: Typical variation in SWATH ship heave response at low speeds as a function of tuning factor. (SNAME)

10.4.1 SWATH hull form teleology

The design of SWATH hulls is largely dominated by the need to design for the transition between platforming and contouring modes. Contouring naturally takes over when wave periods are large, and large wave periods correspond to taller waves. This means that a SWATH will platform in smaller waves, and then when the waves become tall enough to threaten the above-water portion of the ship she will automatically begin contouring those waves.

Designing the hulls for this behavior requires a consideration of the wave heights, wave periods, ship dynamics, and the height of wave at which we want the changeover to occur. This is clear enough conceptually, and under "procedure" I will recap some practical techniques for causing this desired effect.

As background, let me complete the fundamental frequency relationship by reminding the reader of a few points regarding ship motions. Most important is to recall that the Excitation Frequency is Wave Encounter Frequency. It depends upon sea state, ship speed, and ship heading. Figure

10.5 depicts standard wave height / period relationships as established by NATO.

Sea State	Wave Height			Wave period		
	min	mean	max	min	most probable	max
1	0	0.05	0.1	[-]	[-]	[-]
2	0.1	0.3	0.5	3.3	7.5	12.8
3	0.5	0.88	1.25	5	7.5	14.8
4	1.25	1.88	2.5	6.1	8.8	15.2
5	2.5	3.25	4	8.3	9.7	15.5
6	4	5	6	9.8	12.4	16.2
7	6	7.5	9	11.8	15	18.5
8	9	11.5	14	14.2	16.4	18.6

Figure 10.5: NATO Standard sea state definitions

The effect of ship speed depends upon the ship's heading with respect to the seas. In pure head seas, the effect of ship speed is to cause the ship to encounter more waves in a given amount of time, thus to reduce the encounter period for a given wave period. This is depicted in Figure 10.6. The opposite is, of course, true in following seas. The relationship for any given heading can be easily solved using trigonometry.

10.4.2 SWATH hull form parents

With these fundamentals covered, we now turn to look at the choice of parent geometries for SWATH ships. There are several important variations from which we choose:

- Strut configuration: Single Strut v. Twin Strut

- Rudder configuration: Overhanging Struts v. Spade Rudders

For resistance there are three parent hulls available:

- Two high Cp / Low Speed T-AGOS parents

- One Low Cp / High Speed parent

Figure 10.7 illustrates a circular-hull low-speed / high Cp hull, in a single-strut per side configuration, with spade rudders. Figure 10.8 illustrates an alternative low speed hull, again having circular hulls and a single strut per side, but this time with overhanging struts and a different hull volume distribution. Figure 10.9 depicts a high speed SWATH hull, having spade rudders.

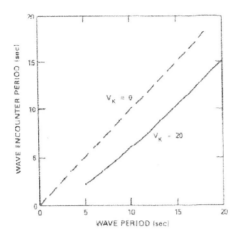

Figure 10.6: FEffect of ship speed on wave encounter period in head seas

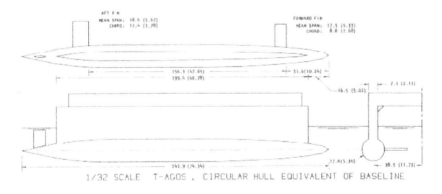

Figure 10.7: High Cp / Low Speed parent SWATH T-AGOS

10.4.3 SWATH hull form development procedure

Having now established the fundamental considerations and lexicon of SWATH hull design, what is a practical procedure to follow to develop such a hull?

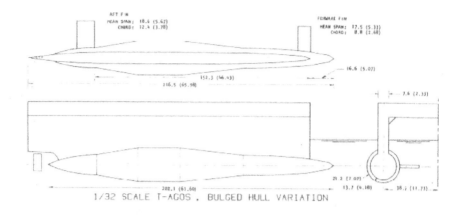

Figure 10.8: High Cp / Low Speed Parent: SWATH T-AGOS-B

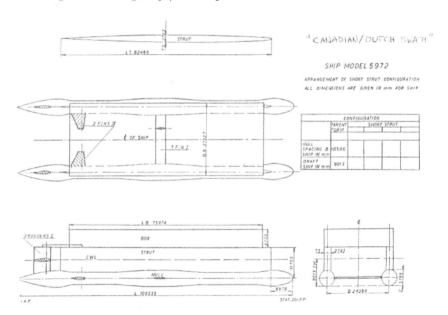

Figure 10.9: Low Cp / High Speed Parent: SWATH 5972

Hopefully the following paragraphs will get one started through the first few turns of the design spiral. For resistance, the only reasonable procedure is a numerical technique such as the previously-introduced Michlet code, the Navy "Chapman" code, or commercial CFD codes. But how do we select the gross hull parameters to feed into these codes, with some assurance that we will have SWATH-like seakeeping? The key to seakeeping design is to select length, volume, and diameter to yield specific target natural frequencies of ship motion. The target frequencies (target periods, actually) are selected by designing for the tuning factor mentioned above. We select the design sea state, and the sea state at which we want the ship to transition from Platforming to Contouring. We calculate the encounter periods (taking account of ship speed and heading) in these sea states. We then design the hulls such that the natural periods are about 0.5 of the encounter period in the design (platforming) sea state, and about 1.5 of the encounter period in the contouring sea state. To accomplish this, we use the following relationships for estimating the natural periods of the ship:

$$T_{HEAVE} = 2\pi \sqrt{\frac{V(1 + A'_{33})}{gA_{WP}}} \tag{10.1}$$

Where:
V =Displaced Volume (m^3)
A_{WP} =Waterplane Area (m^2)
A_{33} = Heave added mass: For elliptical hulls, $A_{33} \approx 0.70$

$$T_{PITCH} = 2\pi \sqrt{\frac{L^2(k_p^2 + A'_{55})}{gGM_L}} \tag{10.2}$$

Where:
L = Ship Length
k_p = pitch gyradius divided by L
GM_L = Longitudinal metacentric height
A'_{55} = Pitch added inertia factor: For elliptical hulls, $A_{55} \approx 0.060$
Gyradius = $\sqrt{I/M}$

$$T_{ROLL} = 2\pi \sqrt{\frac{B^2(k_r^2 + A'_{44})}{gGM_T}} \tag{10.3}$$

Where:
B =Waterline Beam (overall)
k_r =roll gyradius divided by B
GM_T = Transverse metacentric height
A'_{44} = Roll added inertia factor: For elliptical hulls, $A_{44} = 0.20$
Finally, there are a couple of SWATH nuances that bear mentioning: Panama Canal Limits artificially constrain beam. It may rapidly become impossible to attain the desired beam of a SWATH above a few thousand

tons. Of course, this limit will change when the widening of the canal is completed.

Following Seas: In following seas the encounter period may be infinite, or very long. In such cases one may generate low frequency responses that are very large, pulling the wet deck all the way down to the water. Fortunately, the forces involved are modest and are easily overcome with active control surfaces.

Lower hull submergence: SWATH model tests have shown some interesting trends related to the submergence of the lower hull, that it may be possible to exploit in practical design. Consider the two angles Alpha and Beta defined in Figure 10.10 and Figure 10.11. It has been observed that Peak Roll / wave slope =˜ 0.35 x Alpha. Also, Peak Roll =˜ Beta. (This latter means that the ship rolls until the lower hull has risen just to the surface, at which point the roll stops. This seems intuitively logical.)

For a more detailed treatment of this entire subject, see Lamb (Reference [42].)

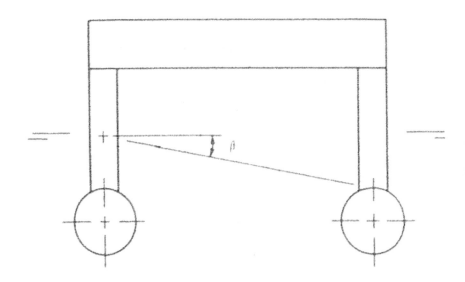

Figure 10.10: Lamb's definition sketch for angle Beta

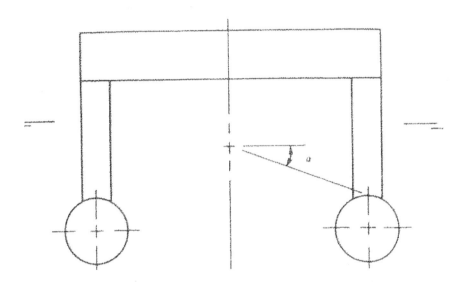

Figure 10.11: Lamb's definition sketch for angle Alpha

11 SWBS 070 - Ship Arrangement

11.1 General Arrangement

"Think INSIDE the box - and outside the cigar."

Catamarans, SES, SWATH, and ACV hullforms all result in ships that are very rectangular in planform as compared to the traditional displacement monohull. Even the trimaran form results in a relatively large cross structure, which doesn't have the "mailing tube" shape of the conventional ship.

For centuries naval architects have had to fit all of a ship's cargo-carrying and living functions into a long narrow railroad car geometry. No longer! The designer of AMVs gets to think "outside the cigar" and inside the box.

Of course, the box-like shape does pose some challenges. Some of these challenges are because our rules - and our clients' expectations - are all shaped by the assumption of "mailing tube" geometry. Consider:

- Corridors - In a conventional ship it is conventional to have a centerline corridor, with cabins giving off the corridor port and starboard, and vertical accesses at various nodes along the length. The monohull passageway network looks something like a fish's backbone. By contrast, the box-like shape of the SES, SWATH, and Cat may suggest that corridors should be in a loop, concentric to the center of the ship itself.

- Outside Cabins - I believe that MSC labor rules require that crewmember cabins be equipped with a portlight. This means that they must be on the outside or perimeter of the ship. Outside cabins are also desireable in passenger vessels. And they are also prominent in the design of luxury hotels, and indeed is one reason that so many big-block hotels are based on internal atrium.

- Distributive System Loops - Again, like the passageways, monohull distributive systems often have a "spinal" architecture. In the case of the "square" ships it may make more sense to base the distributive systems on a horizontal loop concept.

- Access into hulls - The vertical access into the hulls can be a challenge, because it may take most of the available width of the hulls.

- Horizontal accesses below waterline - Fore and aft passage inside the lower hulls may be impossible or difficult, due to the narrowness of the hulls. Also note that naval rules prohibit the installation of doors - even watertight ones - in bulkheads below the waterline. So vertical accesses must be provided, two in every subdivision. This can eat up a LOT of arrangeable area in the narrow hulls.

- Visibility from Bridge - the bridge of any ship must have visibility, including forward, aft, and down (overside). For many AMVs this requirement results in a very wide bridge compared to the length of the ship. But then, the bridge on a 30m wide SWATH is no wider than that on a 30m wide containership, it's just that a 30m wide SWATH is only, say, 10,000 tons, while the containership may be 50,000 tons.

It is difficult to be scientific and mathematical about ship arrangement. Some may claim that this is an area where naval architecture becomes an art. As a result, I think that the best I can do is to take a walking tour of some AMV designs and point out features that are unique about their arrangements, and in which the pioneering naval architect will have to think outside his mailing-tube paradigm, and inside the box.

The first illustration is m/v ANAHI, a Galapagos Islands catamaran tourboat - Figure 11.1. You may be able to see in her configuration that she has devoted the main level of the deckhouse - which is in fact the main level of the ship - to passenger accommodation. Crew and servants are accommodated in the hulls - see the small portholes visible in the topsides. The bridge spans the full width of the deckhouse, including open bridge wings which are presumably fitted with control stations. It's not clear, but I think she also has a flying bridge on top, although I don't know whether this is an operational bridge or a passenger area.

ANAHI describes what may be considered the "generic" AMV configuration, with a rectangular payload compartment sitting atop hulls.

The early USN SWATH "KAIMALINO" (Figure 11.2) not only pioneered the SWATH form, but pioneered the arrangement of a SWATH as well. I am disappointed that we don't see more vessels taking some of KAIMALINO's innovative ideas.

The feature I am most interested in on KAIMALINO is her forward bridge. Note how this forward compartment gives the bridge team unobstructed views forward and down. Note also that the bridge is raised a half-level, so that rearward vision is also provided.

Figure 11.1: Galapagos Islands tourboat ANAHI, showing the standard arrangement of a catamaran

I am also intrigued at the challenges this raised, and how they solved them: Note the two anchors. The anchor chains run up to the top of the bridge, where the windlasses are installed, and the chain lockers are then below decks somewhere. Note also that KAIMALINO shows that the anchors on a multihull don't have to go in the sides of the bows like they are in a tanker. It makes more sense to put them between the two hulls.

The Canadian "PacifiCat" class fast ferry has an innovative bridge solution too. She is depicted in Figure 11.3 and Figure 11.4. Firstly, note that the bridge is not on the topmost deck of the ship. The topmost deck is a passenger lounge - the bridge is one deck below. This means that the bridge team have no aftward visibility and have to rely of CCTV circuits for this, which required a special waiver from Transport Canada.

Also note that the owners insisted on full-width enclosed bridge wings, that overhang the side of the ship. The master wants to be able to stand just past the full width of his ship and be able to see the mooring lines being attached, look down on the fenders and camels, etc. The resulting bridge wings are clearly visible in both pictures, but what is not clear is that the floor of the overhanging wing is plate glass. It is a eerie feeling to stand on a glass plate and look down at the water some ten meters below

Figure 11.2: KAIMALINO, pioneering an unusual arrangement approach

your feet.

RADISSON DIAMOND (Figure 11.8) may be the ultimate incarnation of a box on two hulls. She looks like a hotel - which she is! RADISSON DIAMOND is actually a SWATH, which may be seen in the small sketch reproduced as Figure 11.9. It seems to me that in laying out a ship of this shape, one would in fact turn to a hotel designer more than to a ship designer. Figures 11.5 through Figure 11.7 are pictures of hotels that I gleaned off the internet. I include these as thought-provokers to suggest ways that some designers have arranged a large-volume box.

Note also Figure 11.10 showing the stern of RADDISSON DIAMOND. There are a couple of features of interest in this picture. Not only is her twin-hull shape made obvious, but note also the platform between the two hulls up at the deck. This platform includes a section that can be hydrauli-cally lowered to the water to form a swimming beach for the passengers. Since a SWATH has such low motions, the result is that the ship is an island in the sea, and she even brings her own beach with her.

Another large multihull to successfully embody the square box approach

Figure 11.3: The Canadian PacifiCat fast ferry. The bridge is not the top deck, but the one right below it.

Figure 11.4: A detail of a Pacificat, showing the overhanging bridge wing

Figure 11.5: A luxury hotel atrium. Given the smooth ride of a SWATH ship, why not use a configuration like this?

is the STENA HSS 1500 shown in Figure 11.11. A boxier shape is hard to imagine, although I think the designer has done a remarkable job of making this as good looking as possible. She is a car ferry, with vehicles on the lower cross deck and passengers above them. The bridge is conventional, located in an island superstructure on the top layer. The stacks - in the red-painted area aft - have been kept as low as practicable to minimize interference with aftward lines of sight.

Figure 11.12 shows a ship that is not good looking, but is certainly square - the USN SWATH T-AGOS. To understand the arrangement of this ship we must understand a little of her history: She was introduced to directly replace a line on monohulls. In order to validate the new hull form, the design team retained many features of the monohull one-for-one. Compare the two pictures in Figure 11.13 and Figure 11.14 and note how many pieces are nearly identical, including the stacks, the towed array winch, much of the deckhouse, etc.

Another SWATH is the very small FREDERICK CREED - Figure 11.15 (she is about 80 feet long.) The feature I wished to highlight here was machinery access. CREED's main engines are located inside the bulges in the lower hulls, visible in the picture. They are accessed by ladders in the struts. Now compare the width of those struts, to the width of the men visible in the foreground.

Along this same line, consider the arrangement drawing from INCAT's website, reproduced in Figure 11.16. Look at the lowermost sketch and note the engine placement and arrangement. Even in this large ship the engine room is extremely tight, and providing access and maintenance access to all sides of it can be a real challenge - remember that we went out of our way to make the hulls slender, now we make the ship's Engineer pay the price.

But the INCAT sketch also points out some of the benefits of our box-like shape: The ship is wide enough to turn a car around on deck. A monohull ferry of this capacity might not have room for a U turn.

Finally, consider Austal's sketch reproduced in Figure 11.17. This shows that the flight deck on their ~3000 tonne LCS is the same size as that on a much larger LPD monohull (about 50,000 tonnes), and is much larger than that on a similar-size Frigate (the 4,000 ton FFG.) The point being that the square-box shape may lead to mission utility, in this case much greater aviation facilities, than is possible on a comparable monohull.

Up to this point I have tried to show the very simple "square box" geometry that we are able to use as AMV designers. But I have also shown that there may be some challenges - it's not all beer and skittles. CREED's narrow struts illustrate one set of challenges. Figure 149 - the USN SWATH "SEA SHADOW" may hint at some others.

11.2 Aesthetics

Aesthetics is another topic that is difficult to treat in an engineering course, but I wish to at least venture a few words on the subject.

As engineers we believe that the design ought to be driven by purpose - that teleology ought to dictate our design solutions. What do I mean by teleology in this case? I mean those requirements such as "weight equals buoyancy" or "structural integrity" or "power to match resistance" etc. But where I go further is in my belief that aesthetic purpose is as valid as engineering purpose.

Think about our use of the word "Good". As engineers we agree that a "good" ship is one with the right amount of strength, or stability. These are engineering "goods." But we would also say a ship is "good" if she is beautiful, and I contend that we ought to strive to make them beautiful.

So I offer a few guidelines on what makes a good looking ship. While there is no formula for successful aesthetics, there are three primary "rules" that I think will help a designer start.

Lines of Force: The profile of a ship may be seen to contain some dominant lines that are called "lines of force." Look at the picture of the Fjellstrand 40m Flying Cat VICTORIA CLIPPER IV in Figure 11.20. She has powerful horizontal lines of force, that sweep toward the bow and converge to a point.

Converging stem angles: Notice how CLIPPER's forward lines all seem to converge near the bow - the slope of the deckhouse, the line of the stem, even the rake of the mast combine to put a "point on the arrow" at one clearly perceivable location.

Parallel stem & transom angles: Finally, note how CLIPPER's stern rake matches her bow rake (and is picked up again by her Union Jack paint job.)

These three sets of curves are in harmony on CLIPPER, and make her one of the best looking catamarans I know.

With all due respect, I offer Figure 11.21 as a contrast. She has the parallel bow and stern profiles all right, but her pilothouse interrupts this flow with the forward-sloped windows, and looks like it came from a different ship. She then introduces other lines of force going straight vertically (not parallel to stem or stern) via her tall oval windows, vertical mast, etc.

To end on a positive note, Figure 11.22 depicts STARSHIP EXPRESS, which I again find to combine powerful lines of force and well-harmonized curves, to produce a good looking vessel, despite her basic "square box" configuration.

Figure 11.6: A four-story atrium, with proportions that might fit many AMVs

Figure 11.7: A hotel atrium. Could this be used on a small catamaran?

Figure 11.8: RADISSON DIAMOND, a SWATH cruise ship

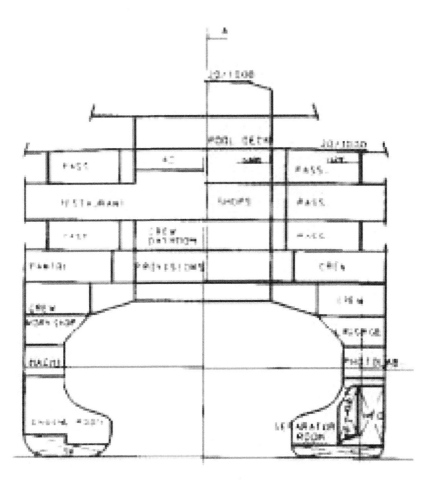

Figure 11.9: A Low-Res section through RADISSON DIAMOND

Figure 11.10: RADISSON DIAMOND Stern View

Figure 11.11: The STENA HSS 1500 fast ferry

Figure 11.12: USN SWATH T-AGOS

Figure 11.13: Monohull T-AGOS

Figure 11.14: SWATH T-AGOS

Figure 11.15: FREDERICK CREED, a small SWATH

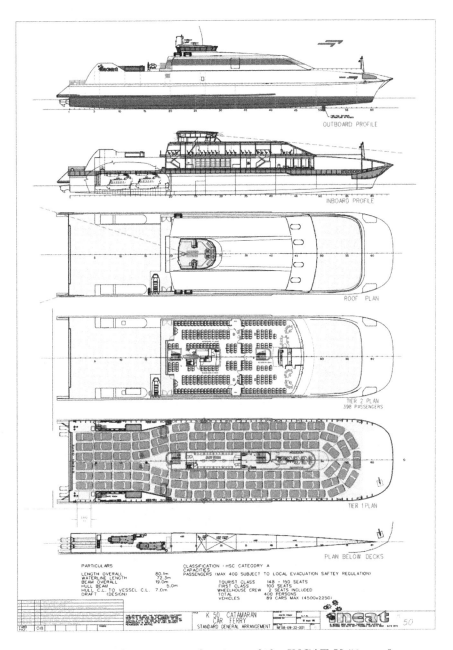

Figure 11.16: Arrangement drawings of the INCAT K-50 car ferry

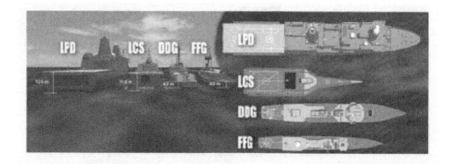

Figure 11.17: Austal's illustration to compare the flight deck size on an AMV versus several monohulls

Figure 11.18: SEA SHADOW

Figure 11.19: SEA SHADOW from above. Note the lower hulls that are dimly visible under the water, forward.

Figure 11.20: VICTORIA CLIPPER IV

Figure 11.21: A counter example, with too many lines going in too many different directions

Figure 11.22: STARSHIP EXPRESS

12 SWBS 079 - Motions & Seakindliness

Here again, the purpose of this work is to present those features of motions and motion control that are unique to AMVs. The student is assumed to have already completed a course in ship motion and to understand the physics of ship motion, and the terminology thereof. In this section I shall devote my attention to how the AMV analysis task is different from that task for a monohull.

The particular areas where AMV motions are unique may be grouped as:

- Operational issues unique to AMVs

- Motions that are themselves unique to AMVs

- Motion analysis criteria need to be adjusted for AMVs

- Motion control options unique for AMVs

12.1 What is Unique About AMV Operations?

The big difference here is that AMVs are designed to specific limiting sea conditions. A commercial AMV is provided with a placard displayed on the bridge that shows the limits for speed and wave height in which the craft can operate. It is the master's responsibility to ensure that the craft stays within those limits.

AMVs *can* be driven outside their permitted envelope, but they *must* not be. Operating an AMV is much more like flying an airplane, or even driving a car, than it is like operation of a displacement monohull: It is entirely possible to go too fast for the conditions, and break the ship, capsize, or otherwise end catastrophically.

Figure 12.1 illustrates an AMV Limiting Condition Table. This happens to be the table for the USN X-Craft, described in Figure 12.2. The figure is quite clear: It says that in waves of height "X", the craft must not exceed certain speeds - e.g. 30 knots in 12 foot seas.

A commercial speed / wave height table will stop at that: It provides simple go / no-go limits for speed and wave height. X-Craft is an experimental ship and her table provides a little more information. Taking again the 12-foot sea case, this table says that below 10 knots slamming is unlikely. Between 10 and 25 knots slamming is possible, and above 25 knots slamming is probable.

Notice that the speed / wave-height table only considers structural limits - i.e. slamming. Slamming is one of the many reasons to slow down in a seaway, but it is the only one that makes it into the table. There is research underway as of this writing to expand the speed / wave height table so that it would be a composite limit including all relevant limiting events, such as human limits, cargo-imposed limits, etc.

Also note that the speed / wave-height table (a) is for head seas, (b) is silent about the wave periods associated with those seas and (c) is for a specific craft displacement.

It seems obvious that craft motions will be different in different headings relative to the sea, and that there should be different slam probabilities and other limits in, say, beam or following sea conditions. It is also logical that there will be differing slam probabilities at lighter displacements, when the cross-structure is further above the water. Finally, the wave periods in shallow water are very different than those in deep water - perhaps the speed / wave height table should be location specific?

I expect that future work will result in a sort of "dynamic" speed / wave height table that takes into account all of these variables, perhaps even reaching the point of being an electronic aid to navigation that is constantly monitoring the ship's limits, against a complex multi-parameter operating envelope.

Why is it important to an AMV designer that there exists a speed / wave height table for his vessel? For two reasons: Because he must design the vessel to the sea conditions in which it will operate, which means that he must know what those conditions are. And conversely, because it gives him another design "degree of freedom" in structural design, in that he may be able to simplify a structural design or reduce weight by changing the operational wave height limit.

12.2 AMV-Unique Motions

AMVs are unique in that they are designed to an explicit wave height limit. They are also unique in possessing a few motions that are not as important on displacement monohulls:

- Corkscrewing

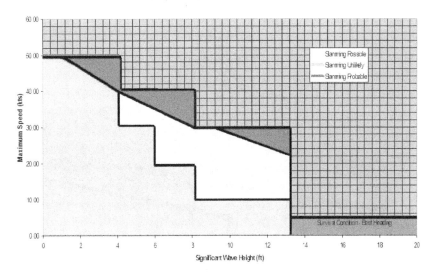

Figure 12.1: The limiting wave height table for the X-Craft, at 1400 tonnes and below, in head seas

- Bow Diving

- Surface Suction

- Cobblestoning

- Plow-In

- Heave Resonance

12.2.1 Corkscrewing

"Corkscrewing" refers to a very uncomfortable motion in which the craft pitches and rolls in a cyclical manner that feels like you are standing on a center-pivoted disk. The path of the body is roughly circular, and it is most annoying.

Corkscrewing is due to the pitch and roll motions being of the same frequency, and maintaining a constant phase relationship. In particular, as a design problem, corkscrewing is caused when the pitch period T-pitch is about equal to the roll period T-roll. This condition is caused, in turn, by

245

Figure 12.2: The X-Craft

the longitudinal and transverse metacentric radii being nearly equal, i.e. $GM_L \approx GM_T$.

The solution to designing to avoid corkscrewing is to design such that the two GMs are not the same.

12.2.2 Bow Diving

Bow diving is a relatively newly-recognized phenomenon, particularly problematic in catamarans. All high speed craft are subject to some undesirable behavior, especially when operating in following seas at a speed nearly equal to the wave speed. In these conditions the craft can surf, or broach, or bow dive. Monohull craft will usually broach, rather than bow-dive, so it is mostly catamarans that experience the dive.

A bow dive can be quite dramatic. Diving the craft so far that there is green water on the pilothouse windows is not unusual. It is also dramatic that this catastrophe comes on suddenly - there is no gentle build-up: One moment your are doing just fine, and then "wham!"

The UK MCA has recently completed a study of this phenomenon [43], and I will usually show the MCA guidance video in class to illustrate the situation. The MCA found no operational guidance better than to simply slow down, and avoid operating at or near wave speed in waves of the same length as the craft. Figure 12.2 provides a graph that shows the deep-water relationship between wave speed and wave length. For a 40m craft avoiding undesirable behavior in waves of 40m length, you can see that this suggests avoiding a speed of 15-20 knots, which is just the speed that one might be tempted to seek as seas get rough.

The MCA did not study SES craft or Trimarans, and it is not known whether these types are susceptible to bow diving in the same way. They did study a small range of design parameter variations, and they found unsurprisingly that increasing freeboard forward is the best way to avoid a bow dive. But this conclusion is qualitative only, and there is no clear guidance on how much is enough.

Figure 12.3 reproduces an MCA photo sequence of the history of a bow dive with a catamaran model in a towing tank.

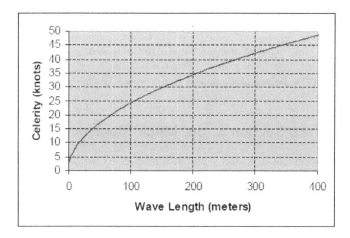

Figure 12.3: The relationship (in deep water) between wave speed (Celerity $= \sqrt{(gL/2\pi)}$) and wave length

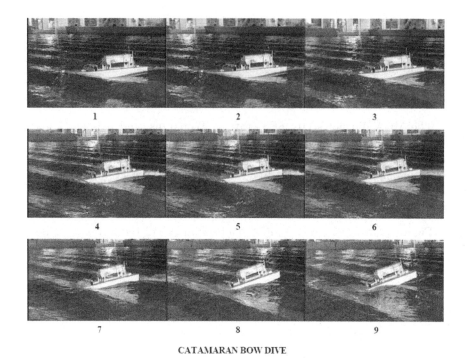

CATAMARAN BOW DIVE

Waves 10° off stern, 6.7% waves, 110% wave speed

Figure 12.4: MCA Photo sequence of model tests of a catamaran bow dive

12.2.3 Surface Suction & the Munk Moment

This next motion class is one I am ill-qualified to lecture on, but which I must in good conscience at least mention. These two forces - the surface suction force and the Munk moment - cause destabilizing behavior for submerged bodies.

Surface Suction is a force that affects a submerged body travelling close to the surface, like a shallowly-submerged submarine or torpedo. In this condition, despite the body being aimed straight and true, the forces are not symmetric - there is a net upward force caused by the free-surface pressure condition, which causes the submerged body to "broach" to the surface. This is why toy torpedoes will "hop" to the surface when towed on a string (as in low-budget movies.)

The Munk moment is similar, but is an effect due to yaw or pitch. It was discovered during the study of airships (buoyant aerostatic craft.) In this case by introducing pitch on a submerged body the pitch causes a pitching moment in the same direction, i.e. tending to exacerbate the pitch. This force again will result in a body hopping to the surface, or plunging to the depths.

Both of these forces may become important in SWATH design. Indeed, in early days of SWATH development I observed a few model tests specifically to characterize this behavior, and the general conclusion was "there's nothing you can do about it, just make sure you put control fins on the SWATH."

That advice remains sound, at least as a starting point.

12.2.4 Cobblestoning

Cobblestoning is a motion type that is unique to SES. According to Yun & Bliault [8] it is not fully understood theoretically. It appears to be a compressibility effect within the air mass of the cushion. This makes the cushion air bubble into a high-frequency spring, and the result is a high-frequency vibration that feels like a car on a cobblestone street - hence the name.

Cobblestoning is addressed only by including active pressure controls on the cushion. These controls consist of louvered vent valves which may be opened or closed to release or retain cushion pressure. Computer-controlled, they end up chattering at a very high frequency ($\tilde{}$ 10 - 100 Hz) to attenuate the pressure spikes from cobblestoning.

It is interesting to note that cobblestoning is not present in very small SES, such as towing tank models. In fact, important work by Steen and Faltinsen [44] showed that the motions of ride control equipped full-scale SES were well represented by the motions of an un-equipped model. Apparently the model scaling issues exactly compensate for the lack of the RCS attenuation of the cobblestones. Thus the journeyman SES designer should leave room in his budget to contract with a ride control supplier.

12.2.5 Plow-In

Plow-In is another motion that is unique to ACVs, and possible with SES. It is described as the skirt tucking in under the craft, and then the craft pitching downward and plowing forward, potentially even capsizing in a pitch-pole event.

Plow-In arises due to lack of pitch restoring force in cushion-supported craft It is driven by added drag of added wetting of the bow skirt. See the diagrammatic sequence in Figure 12.5: The skirt begins to tuck under, the skirt drag increases, the drag down low causes a bow down moment, the bow down attitude means that more skirt drag arises, the skirt tuck-in causes the center of pressure to move aft which causes further bow down trim, etc.

The whole sequence can be very fast and can lead to results including capsize or pitchpoling.

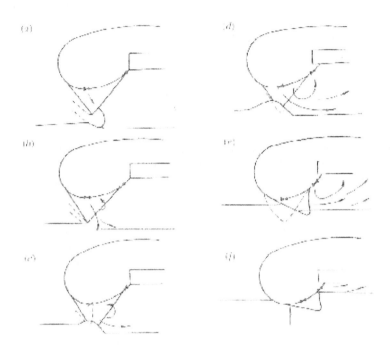

Figure 12.5: The Plow-In process, from [8]

12.3 AMV Motions Analysis & Criteria

In the preceding sections I have stated that AMVs are explicitly designed for a certain wave environment, and also that AMVs may experience some unique motions, which do not trouble their displacement monohull cousins. In consideration of these facts, there are also differences in the way that AMV motion analyses should be conducted. But let us begin by reviewing the manner of motion predictions for conventional ships.

In general, the best procedure for a comprehensive seakeeping assessment of a ship is that given in NATO Standardization Agreement STANAG 4154

[45]. This method, greatly summarized, is as follows:

1. Establish a list of missions.

2. Establish mission-based criteria sets = motion limits.

3. Perform motion predictions

4. Compare motions vs. limits at each speed/heading combination, for each sea state of interest

5. Calculate Operability Index (OI) for each sea state of interest

6. Calculate overall Seakeeping Performance Index (SPI) by applying probability of each sea state.

In addition to describing this methodology, the STANAG also offers some preliminary suggestion as to what motion limits to use when assessing the pass/fail criteria.

As a result of this approach, there is a large importance in the choice of the mission-particular motion limits.

12.3.1 AMV Motion Criteria

The motion limits for a mission are derived from analysis of the characteristics of the equipment essential to that mission: Engines, cranes, missile launchers, people, etc. In many cases it turns out that the motion-critical piece of equipment is the ship's crew. This makes sense, since all the man-made components were designed for a ship's motion environment, whereas we do not have the luxury of redesigning the humans. Because of this, I will discuss the motion limits as if the human system is the limiting system on the ship, but it is important to realize that this discussion can be extended to all of the motion criteria in common use.

The crux of the problem with the STANAG motion limits, and other similar ones, is that many roll & pitch limits are actually surrogates for lateral accelerations. For example, the STANAG motion limit for seasickness is 8° of roll and 3° of pitch. It is not, really, true to say that humans get sea sick if we simply tilt them sideways 8°. Nor that they get seasick if we lean them forward 3 degrees. Indeed, why are these limits different? Is the cure for seasickness simply to turn and face athwartships? The 8 degree roll limit is really only true if that roll is roughly sinusoidal, and has a period of about 10 to 20 seconds. In fact, a more reliable method for predicting human performance degradation is to use the Motion Sickness Incidence (MSI) and Motion Induced Interruptions (MII) calculations.

The MSI method uses O'Hanlon & McCauley frequency-domain criteria, from Reference [46]. To use this method one calculates overall vertical motion response spectra, and compares these to published threshold spectra.

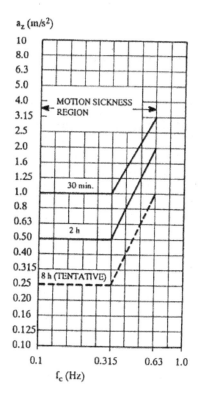

Figure 12.6: O'Hanlon & McCauley criteria for motion sickness, as presented in ISO 2631

MII is calculationally much more complex. It calculates the likelihood of "jarring" a person who is trying to perform a task, and in doing so it captures the input from impulsive events as well as regular sinusoidal motions. The MII analysis captures the horizontal plane motions - including those which arise from vertical plane ship behavior.

I suggest that the use of STANAG-like pass/fail motion limits is not appropriate for most AMVs, and that an MSI / MII analysis is a more realistic assessment of ship operability.

12.3.2 Added Resistance

Added resistance is not a motion, but it is a wave-induced effect and so I have included it here.

For most AMVs added resistance is due to the same hydrodynamics that we learned in monohull courses. For powered-lift craft, the best I can offer in this area is to quote Faltinsen [4] for a semi-displacement vessel: "[The added resistance due to waves] *is caused by diffraction of the incident waves by the ship and by radiation of waves due to wave-induced ship motions. A dominant effect for an SES is associated with the leakage from the air cushion caused by the relative vertical motions between the SES and the waves. If the lifting power of the fans for the cushion is unchanged, the air cushion pressure drops and the SES sinks to a lower position with a larger wetted surface. The calm water resistance in this lower position explains the major part of added resistance for an SES in a seaway.*"

In other words, for an SES much of the added resistance is due to increased air loss leading to lower effective cushion pressure, and thus operation on a higher resistance part of the performance space.

12.4 Motion Control for AMVs

Up to this point I have focused upon wave-induced motions. In doing so, however, I have already acknowledged the existence of ride control (motion control) systems. Let me also acknowledge that there are motions we want to induce even in calm water, in order to steer the craft. What are the modes of control available to us? What types of motion control devices are employed on AMVs? How effective are they?

12.4.1 Modes of Control

There are three ways of control a motion. Sometimes we just clamp the moving object in place, restricting it's ability to move. But there may be other interventions earlier in the chain of events that turn out to be more effective. When dealing with an oscillatory motion, such as ship motions, we can:

- Eliminate the excitation

- Damp the excitation / transmission pathway

- Counter the force directly with an anti-force

Most motion control devices take the third approach: Counter the force directly by creating an equal and opposite force. But note that, with AMVs at least, there are some opportunities in the other two areas. Very fine waterplanes, as used on a SWATH or even on an SES, might eliminate the excitation, meaning that there is no motion to resist. In the case of an SES there is also a waterplane area due to the cushion, and keeping this exciting force small requires cushion pressure ride control.

The second approach "damp the transmission pathway" I have rarely seen used. One craft that does this is Ugo Conty's "spider boat" W-AMV (Wave-Adaptive Marine Vehicle) which uses a suspended gondola, such that motions induced on the craft's hulls never make their way to the crew cabin. This unique ship is pictured in Figure 12.7.

12.4.2 Effectors

What force-producers are available as ship control effectors? I will describe seven classes of device, and attempt to describe the capabilities, attributes, and effectiveness of each of them.

- Cushion-based ride control

- Foil-based ride control

- Interceptor-based control devices

- Propulsor steering (e.g. waterjets)

- High speed rudders

- Aerodynamic Steering & Control

- Cushion-Air Thrusters

Cushion-based ride control

I mentioned cushion-based ride control a few paragraphs above, when discussing cobblestone motions. On an air cushion supported craft there is some 70-100% of the craft weight that is borne on the cushion. This means that the cushion can act as a vertical force generator able to produce 0.7-$1.0g$ accelerations on the craft. This is a very powerful force, if it can be controlled and used.

One issue to consider in using cushion-based ride control, is the speed with which the forces can be generated. In current practice air cushion ride control systems involve computer-controlled cushion vent valves, driven by a heave-minimization algorithm. The algorithms are proprietary. The two

Chronicle / Paul Chinn

Figure 12.7: Ugo Conti's Spider Boat. Photo from SFGate website.

producers of SES RCS that I know of are Maritime Dynamics in Lexington Park, MD, and Island Engineering, in nearby Piney Point Maryland.

The use of vent valves for heave control is fast, but it us inefficient - it requires that the craft's lift system be oversized so that air capacity is available for some to be "dumped" in response to pressure spikes, and then the fan must rapidly resupply that air when the valves close. We will discuss the dynamics of lift fan operation in a later chapter, but this simple "dump / refill" mental model will suffice at this point.

An alternative to dumping the air is to actually throttle the fan somehow. Now, throttling the fan engine is not practical, because a ride control system needs a response time of .01 second. But it is possible to throttle the fan aerodynamically, by installing louvers or guide vanes on the inlet side of the fan. Choking the flow on the inlet side will change the fan's pressure-vs-flow characteristic, and can act as a ride control effector. In practice however there are very few inlet-side SES RCS systems - the vast majority of SES use vent valves.

Note that a simple SES will get only heave control from a cushion-control system. There is no significant roll or pitch force generated by changing the cushion pressure. Many people have experimented with ways to change this by fitting intermediate skirts or longitudinal divider skirts. These techniques work, but it is very difficult to create the intermediate skirt in a manner that has acceptable drag.

Foil-based ride control

Many craft use foil-based ride control systems. Of course, most of the hydrofoil craft do this, but a very large number of SWATHs, catamarans, and trimarans do too.

The foils used are simple rudder-like structures, oriented and actuated to produce a resultant force in the desired direction, at the necessary location on the ship. Roll control foils (anti-roll fins) are common on monohulls. Pitch control foils, such as those made by Maritime Dynamics and depicted in Figure 12.8, are fairly common on catamarans.

The drag of the foils is calculated by conventional techniques, even techniques used for displacement ship rudders, and must be added to the drag of the ship.

Interceptor-based control devices

T-Foils are commonly used near the bow as pitch effectors. Near the stern it is more common to use interceptors.

An interceptor is effectively similar to a trim tab, which many readers are already familiar with. Maritime Dynamics provides, on their website,

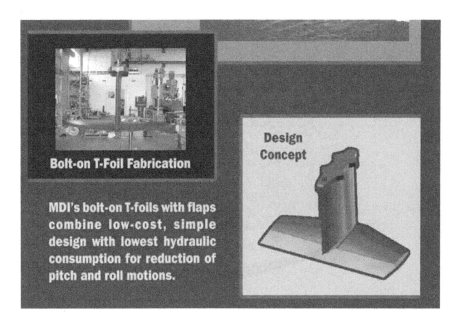

Figure 12.8: A Maritime Dynamics T-foil

the four illustrations reproduced as Figure 12.9 though Figure 12.12 which compare and contrast Trim Tabs with Interceptors.

The interceptor is a guillotine-like blade that intercepts the flow of water close along the hull. This causes a sudden rise in pressure forward of the "obstruction", and this pressure results in a lift force acting on the bottom at the stern.

Interceptors appear to be highly efficient, having high lift-to-drag ratios in most cases (perhaps L/D=10 or more.)

Propulsor steering (e.g. waterjets)

All of the effectors described thus far have been for motion control. What about steering?

Many AMVs are fitted with waterjet propulsors, and these are used for steering as well. Note however that propulsor steering need not be unique to waterjets: Recreational craft with outdrives or outboards are also propulsor-steered.

Propulsor steering is highly effective. The side force generated is simply the propulsor thrust times the sine of the steering angle. If we assume that a typical AMV has a Lift to Drag ratio around 10:1, and a steering angle

257

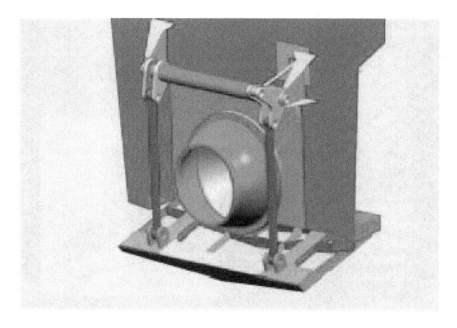

Figure 12.9: An MDI Trim Tab, 3-D view

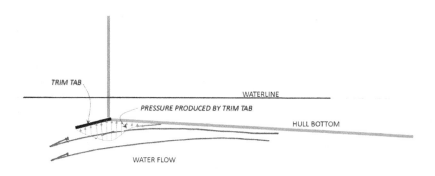

Figure 12.10: A trim tab, profile view, showing the pressure effect on the bottom.

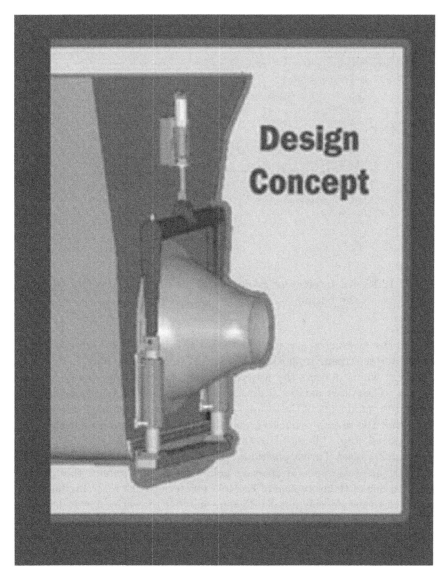

Figure 12.11: An MDI Interceptor, 3-D view

around 30 degrees, then this results in a side force equal to 1/20 the weight of the ship. Unlike a rudder, this force is available at any ship speed (if top throttle is used) whereas a rudder's force varies as V^2.

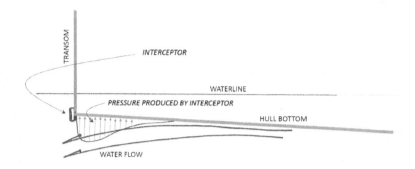

Figure 12.12: An Interceptor profile view, showing the pressure effect on the bottom.

Waterjet steering is accomplished via a steering "nozzle" which deflects the waterjet stream from side to side. Waterjets also have a reversing "bucket" which deflects the waterjet stream forward, to generate reverse thrust. These two devices comprise a system, but they are usually two separate components of the system - one for steering and one for reversing. Note that the reversing buckets are "throttleable," meaning that they can be adjusted from full-ahead to full-reverse, including a 50%-ahead/50%-reverse=Zero-Net-Thrust position.

There are two types of steering and reversing assemblies that I have seen. In one of them, typical of KaMeWa waterjets, the reversing bucket is attached to the steering nozzle. This means that the entire flow is deflected through the steering angle, and then some fraction of the flow is directed backwards as reverse. This results in steering vector as shown in Figure 12.13 - the net waterjet resultant is a single vector of some magnitude, deflected at the steering angle.

The other type of steering / reversing arrangement uses a bifurcated "rams horn" duct for the reversing bucket. This duct captures the waterjet outflow and then redirects the captured portion forward in two streams, aimed slightly port and starboard of centerline. This type of duct is used on HamiltonJet waterjets. The steering nozzle is forward of the reversing duct, and directs the waterjet outflow from side to side. When steering and reversing simultaneously, the steering nozzle delivers more flow to one "horn" of the rams horn than to the other. This produces a complex mix of vectors as depicted in Figure 12.14.

One impressive feature of this style of steering / reversing suite is that it can be throttled to produce a resultant vector that is perfectly sideways, with no fore-and-aft component. Proving this fact will be assigned to the student in a homework set.

Finally, and forming the second aspect of the mentioned homework set, with any type of steerable propulsor, including outboard motors and out-drives, a widely spaced multihull can vector the thrust and the steering angles such that the result is a pure sideways thrust *through the ship's center of gravity*. This means that the ship can move sideways from the pier, with zero headreach and zero rotation.

This capability is due to the combination of the steerable thruster and the wide beam of the ship. It also explains why very few multihull AMVs are fitted with bow thrusters.

The waterjet steering nozzle does reduce the net thrust of the waterjet. This is due both to the steering itself, wherein the thrust vector is diminished by the cosine of the steering angle, but also due to hydrodynamic drag due to friction on the inner walls of the steering nozzle. There is a lot of very high speed water moving through the nozzle, and the drag forces (and the mechanical loads on hinge pins and actuators) is substantial.

During early days some waterjet manufacturers tried to avoid this loss by making their steering buckets slightly larger in diameter than the waterjet plume. As a result, when in the dead-ahead position the nozzle didn't touch the flow, and this drag was eliminated. These manufacturers could claim a few percentage points higher thrust than their competitors.

Unfortunately, this also meant that the nozzle didn't produce any steering effect until it had been deflected those few degrees needed to bring it into contact with the plume. This resulted in a dead band in the helm that was very annoying to operators.

High speed rudders

One way to avoid the cosine-loss effect due to propulsor steering is to es-chew propulsor steering and fit rudders instead. The problem of course is that rudders may be draggy. Some architects have therefore struck an interesting compromise, fitting very small rudders that are only suitable for course-keeping at high speed, and switching to propulsor steering at low speed or when large track deviations are needed. The switchover is handled automatically by computers in the control system.

High speed rudders of this sort are often fitted upstream of the propulsor. This is obvious when speaking of waterjet driven craft, but less obvious

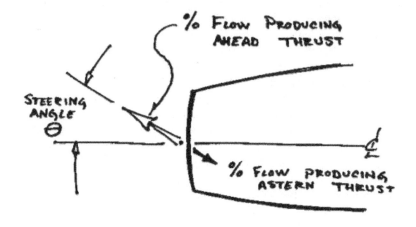

Figure 12.13: The steering forces due to a KaMeWa-style steering and reversing suite

with propeller driven craft. However, if the propellers of choice are fully-ventilated surface-piercing propellers (which are very high efficiency devices with many superiorities over waterjets) then upstream is about the only reasonable place to put a rudder - because downstream of such a propeller the flow is too mixed with air and too energetic to be used efficiently. As a result it is not uncommon to find small nearly rectangular planform rudders protruding below the keel on high speed craft.

Other rudder solutions have been experimented with. One intriguing one was the "plunging rudder" used in some early INCATs. A conventional rudder is a "wing" of fixed geometry and variable angle of attack. A plunging rudder was a retractable wing with fixed angle of attack. Two such rudders were fitted, one on either hull of the catamaran. They were oriented oppositely, say "toed in" on either side. Then, to turn to starboard you lower the starboard rudder. To turn to port you lower the port rudder. The amount that the rudder is lowered determines the amount of steering force generated. The perceived advantage was that when no steering was required, there were no rudders in the water and thus no appendage drag.

An interesting homework assignment will be to calculate the comparison between the drags of these two rudder concepts.

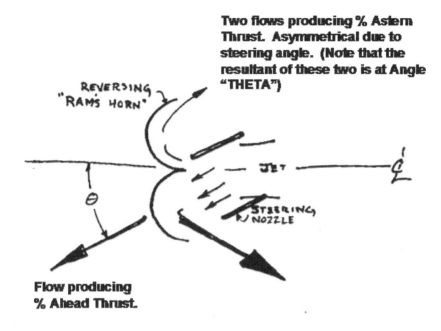

Two flows producing % Astern Thrust. Asymmetrical due to steering angle. (Note that the resultant of these two is at Angle "THETA")

Flow producing % Ahead Thrust.

Figure 12.14: The steering forces due to a Rams-Horn style steering and reversing suite

Aerodynamic Steering & Control

All of the above steering devices generate force by acting on water. But some AMVs - especially the amphibious hovercraft - generate their steering and control forces aerodynamically. This is true for both their rudders and their bow thrusters.

Figure 12.15 depicts an LCAC. Figure 12.16 shows a blow-up of that photo, focusing upon her propulsion nozzles and the airplane-style rudders that are located in the slipstream behind them. This design is straightforward and within the skills of any naval architect who remembers to change the density of his working fluid.

Figure 12.17 shows another detail of the LCAC, showing her cushion-pressure bow thrusters. These are best described with reference to the next paragraph.

Figure 12.15: An LCAC Class ACV

Cushion-Air Thrusters

Consider a SES or hovercraft having an air cushion pressurized to 200 psf. If we open a 4'x 4' "door" into that cushion we will experience a force of 4x4x200=3200 pounds. This is about equal to the thrust of a 100-hp marine bow thruster.

This technique is used quite effectively on SES and ACVs. Figure 12.17 shows the steerable thrusters on the LCAC ACV. These "snorkels" can be rotated through 360 degrees continuously to give thrust in any direction. They are used in the LCAC to provide needed side thrust for coursekeeping and close-in maneuvering, especially needed since the LCAC's do not have steerable propulsors.

Similar function can be attained on an SES if the ride control vents are placed in the sides of the ship. Manual override of the Ride Control System can open these vent valves, and produce a corresponding thrust at their location. This can be used to side step away from the pier, or to hold the ship onto the pier so lines can be passed.

Note that if these vent valves are above the height of the pier, they will also deliver a hurricane of wind to any spectators or line handlers who are present! Their position and placement may need to take this into account.

Figure 12.16: A blow-up of the LCAC's propulsion nozzle, with the rudders marginally visible behind them.

(i.e. don't place them too close to the mooring stations.)

As a final comment on the control effector suites that have been used on AMVs, I offer a picture of the appendage suite on the 1970s experimental testcraft SES 100A. In addition to the ones labeled, note the complex stability planers that are visible forward on the starboard side. Each one of those appendages was there to solve a particular need, and a student of AMV design would do well to ponder the result.

Figure 12.17: A blow up of the LCAC's bow thrusters (the snorkel-like structures near the center of the photo.)

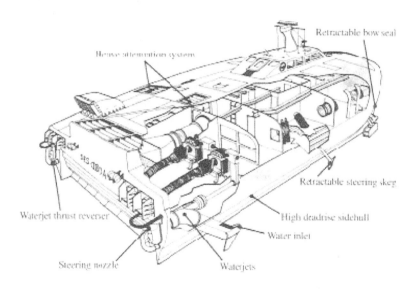

Figure 12.18: The many appendages of the SES 100A

13 SWBS 079 - Stability

What is unique about AMV Stability? Very little is unique about the physics, but we find the particular resulting stability curves may be a little surprising, the criteria may need to be special, and the measurements may be difficult.

13.1 Stability Curves for Multihulls

In this section I wish to highlight some of the "surprising" features of the stability curves of the most common AMVs. I shall address:

- Catamarans

- Trimaran

- SWATH

- SES

In most cases AMV stability is the same as monohull stability. For powered lift craft, the air cushion has a de-stabilizing effect, which can be important to SES and ACV. But let's begin with a sort of "refresher" look at a monohull stability curve. Figure 13.1 depicts a righting arm curve for a generic monohull with circular sections, for the case where G is below B. In this simplified case it is the lever arm GB that yields the vessel's righting arm, and the righting arm curve is a simple sinusoidal shape.

For the more common ship case where G is above B, the shape of the righting arm curve depends upon the transfer of the center of buoyancy, as the vessel heels - so-called form stability. Figure 13.2 depicts a generic righting arm of this sort. The curve is still roughly sinusoidal, but not mathematically so. It rises gradually to a peak somewhere in the range (usually) 45 - 90 degrees, and then has a second zero crossing somewhere beyond about 100 degrees. The slope of the curve at the origin - $\frac{dRA}{d\Theta}$ at $\Theta = 0$ - is the GM.

Figure 13.3 is a generic sketch of a trimaran righting arm curve. What happens in this case is that the immersion of the amas causes a greater shift in center of buoyancy than would otherwise be possible for such a narrow hull. But despite the unique shape of the hull, the shape of the curve is still quite monohull-like. That is, up until the angle at which the amas are fully submerged (or fully emerged) in which case only the stability of the narrow main hull remains.

Now let's consider a catamaran. Figure 13.4 depicts a catamaran righting arm curve. Note the fact that the peak of the curve occurs at a very low heel angle. Why is this? Consider the shift in the center of buoyancy: At the instant that one hull lifts clear of the water, the CB is fully located at the other hull. From this angle onward the righting lever diminishes in the shape of a cosine. Thus the maximum righting arm is very early. Note also that the angle of zero crossing may be less than 90°, and finally that the stability in the inverted position is nearly the same as the stability upright. All of these features are well known to sailors of recreational catamarans!

This boils down to saying that the trimaran has a stability curve that is generally like that of a monohull with G above B, while the catamaran has a stability curve that is rather different. The catamaran curve is marked by a very high GM, but this high GM does not mean that there is a lot of stability - the actual area under the righting arm curve may be modest, depending on where the zero-crossing is located.

13.2 Hydrofoil Stability

This section is very incomplete - indeed, almost nothing more than a placeholder.

The problem with hydrofoil stability is to find a means of generating a restoring force. Consider a fully submerged foil system, and sketch the free body diagram of the system. Now sketch the FBD of the system with some arbitrary roll angle. You will quickly see that there is no restoring force generated, as long as the foils stay submerged.

Solutions to this for fully submerged foil systems are to rely on active control.

Passive solutions rely on changing the amount of foil lift as the vessel rolls. This is done by having foils that cross the water surface, so that they lose foil area (lose lift) on the high side, and gain lift (gain area) on the low side. This results in a roll restoring moment.

Note: Now expand this discussion to consider the pitch / heave stability of a hydrofoil - it's even worse! Now not only is there no restoring force, but the pitch forces are unstable - increasing the pitch angle increases the lift which increases the heave and eventually the boat flies out of the water! How shall we solve that?

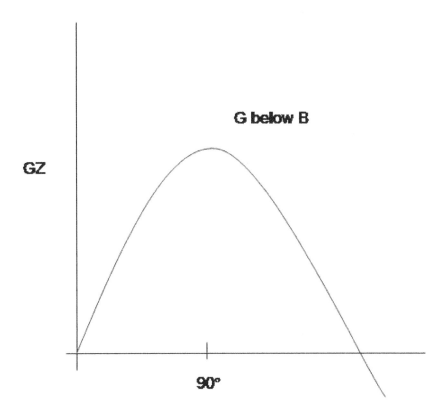

Figure 13.1: Monohull Stability - G below B

13.3 SES Stability

Surface Effect Ships, due to the presence of the air cushion, have their own stability novelties and deserve separate treatment.

13.3.1 SES Static Stability

First, let us consider the static case of stability at rest. The pioneering work in this area was done by Mr. Andrew Blyth and published by RINA (Reference [47].) Blyth's illustration of an SES midsection is reproduced in Figure 13.7. As may be seen, the pressure due to the air cushion has a destabilizing effect - the upward vector representing the powered lift is

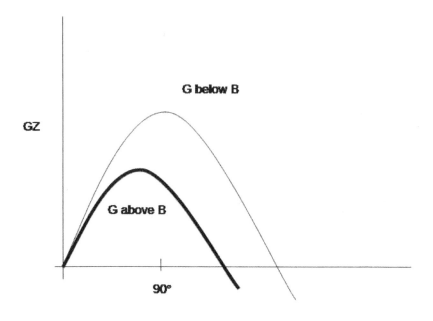

Figure 13.2: Monohull Stability - G above B

located on the upsetting side of centerline, and (think about the trigonometry) it moves further off center as the angle increases - it becomes more upsetting.

The hydrostatic part - the righting moment caused by the waterplane area - is given by the same integration of waterplane inertia as seen with all other ships. What is special in the SES case is the need to add the destabilizing effect of the cushion. Blyth resolves these forces and has published the following formula for GM of an SES:

$$GM = I_T/Vol + drafts S_c(P_c^2)/Vol - KG \qquad (13.1)$$

Where:

I_t = Sidehull Waterplane Inertia (both sides)

S_c = Cushion Area

P_c = Cushion pressure head (meters)

KG = Height of Center of Gravity

Vol = Immersed volume = $\Delta/\rho g$

Note in the formula that the effect of cushion pressure is of the form $\frac{x}{Vol}$. In fact, it is a loss of waterplane inertia exactly analgous to a free

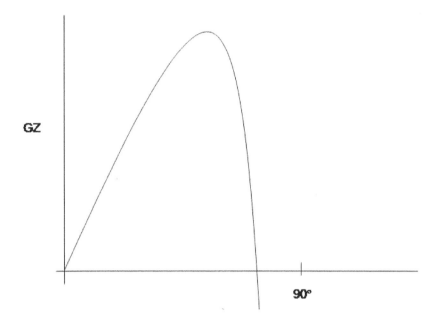

Figure 13.3: Trimaran Stability - G above B

surface correction. The waterplane inertia of the SES sidehulls consists of the inertia about the sidehull's own centerline, plus an Ad^2 term due to the shift of parallel axes to the ship's centerplane. In most cases the Ad^2 contribution is larger than the sidehull inertia about it's own axis, such that the smaller term can be completely neglected. Considering this fact, Yun & Bliault [8] have published an approximation formula, as follows:

$$GM = \frac{\rho g A_s (B_c + \frac{A_s}{L_s})^2}{2W} - \frac{\rho g S_c P_c^2}{W} + \frac{L_s tan(T)}{2} + P_c - KG$$

Where:
$A_s =$ Sidewall Waterplane Area (one side)
$B_c =$ Cushion Beam
$L_s =$ Sidewall Length
$W =$ Craft weight (N)
$S_c =$ Cushion Area
$P_c =$ Cushion pressure head (meters)
$T =$ Trim Angle
$Vol =$ Immersed volume $= \Delta/\rho g$
As may be seen, this formula replaces the waterplane inertia with an

271

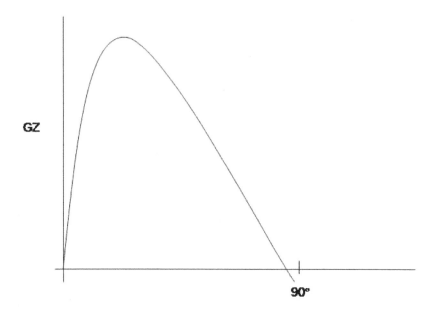

Figure 13.4: Catamaran Stability - G above B

approximation, and most of the other terms are the same as Blyth's. I admit to being puzzled as to where the trim correction came from.

Blyth's formula shows the destabilizing effect of the cushion. Without an air cushion the SES would have a stability curve based on catamaran hydrostatics. And, when it is on-cushion, there comes some angle of heel at which the cushion escapes and becomes ineffective. (When the high-side keel comes clear of the surface, it is no longer possible to have any sort of cushion.) Above this angle of heel of course the vessel is no longer an SES and is a catamaran.

Blyth illustrates this transition in the drawing reproduced as Figure 13.8. In this drawing he has two curves, representing the SES on-cushion and off-cushion. Above some critical angle the on-cushion curve "goes away" and only the off-cushion case exists.

Blyth's drawing is also helpful in two other areas: Note that it very clearly shows the reduced righting energy (area under the curve) that is due to the destabilizing effect of the cushion. Also notice the very shallow slope of the curve at $\Theta = 0$: The GM of the ship is very small, perhaps even zero. This makes sense when we consider the very small waterplane,

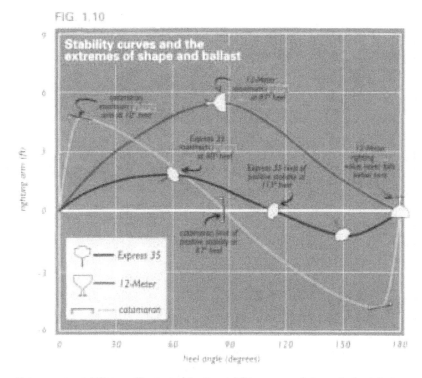

Figure 13.5: Taken from a forgotten site on the internet, this graphic does an excellent job of contrasting the stability of three types of craft.

and the importance of the negative cushion term, finally coupled with a normal "G above B" mass property.

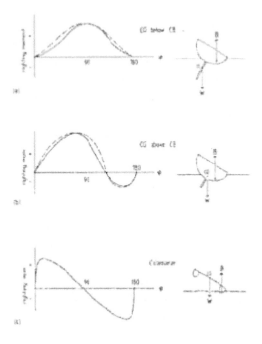

Fig 5.9 Three generic forms of the static stability curve. In
(a) the centre of gravity CG is always below the centre of
buoyancy CB. At all angles of heel the hull has a positive
righting moment tending to bring it back upright. The
dashed line shows the modification that would occur for
increased beam (b) shows the situation where the CG is
above the CB, which is common in many modern designs.
There is a region of negative righting moment, implying that
the boat has some stability in the upside-down position. The
dashed line shows the effect of increased beam. A more
extreme case of this kind of curve is shown in (c). A
catamaran relies almost entirely on form stability. The
symmetry of its hulls produces complete symmetry of the
righting moment curve, where it is seen that the boat is just
as stable upside-down as right way up.

Figure 13.6: Another internet-harvested graphic, depicting the situation.
The condition of a trimaran is like that of a monohull with G
above B.

S.E.S. INITIAL STATIC STABILITY

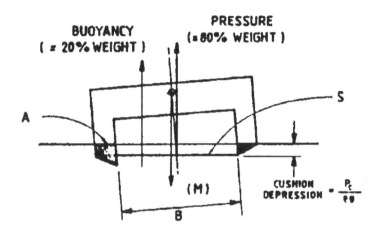

Figure 13.7: Blyth's illustration of the balance of righting forces for an SES on cushion.

13.3.2 SES Dynamic Stability

The above section addressed the stability of an SES at rest, both on- and off-cushion. When forward speed is introduced there are additional forces that come into play that may have an effect upon the transverse stability of the craft.

The following lecture material is derived from notes given to me by John Lewthwaite, while he was working for the German Ministry of Defense in 1986. Some of this material was then amplified by Andrew Blyth, working for the UK MCGA in the same time frame (Reference [48].)

Figure 13.9 depicts the stability-related forces acting on a typical SES whilst in a high speed turn. The buoyant, cushion pressure, and gravity forces are the same as those when at rest. Due to the speed and the turn we have to add a centrifugal force trying to roll the ship outward, a rudder force trying to roll the ship inward, and a planing force which may go either way.

The planing force is considered to act normal to the deadrise surface of

275

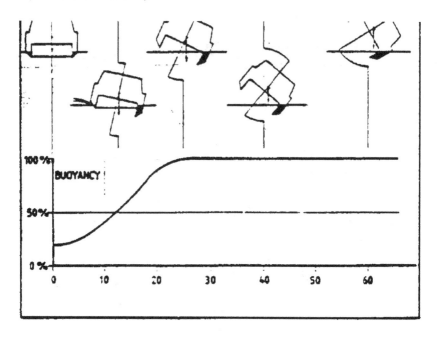

Figure 13.8: Blyth's illustration of the effect of emergence of the sidehull as
an SES heels

the hull. In the case illustrated the planing force passes above the CG of
the craft, with the result that this force acts to roll the craft inward in the
turn. If the CG were higher then this force would pass below the CG, and
would tend to capsize the craft in the turn. The same would be true if
the deadrise angle were much higher. Indeed, the importance of this was
found at great cost when an early vertical-sided SES testcraft rolled over
in a turn, killing the test pilot.

The design of the planing surface angle, to ensure that the force vector
passes above the CG, is one of the major design drivers in deciding the SES
beam and deadrise angle, as discussed under hull form design.

In most design projects these forces are not actually calculated in detail. Instead it is generally sufficient to design the craft for adequate static
stability, and then to design the sidehulls to ensure that the planing force
resultant passes above the CG. The total success of the design is then val-
idated in free-running model tests, in waves.

There is a further subtlety of this planing force stability, which is its
own instability with respect to roll angle: As the craft rolls outward, the
planing force moment becomes increasingly a destabilizing moment. This

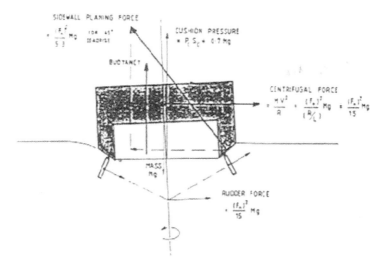

Figure 13.9: Forces acting on an SES in a high speed turn

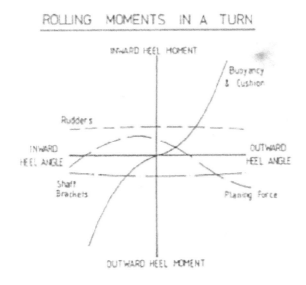

Figure 13.10: The roll moments associated with the forces in Figure 13.9

is depicted in Figure 13.11. The uppermost illustration in that figure shows the un-heeled case. Directly below is a case with the craft heeled inward. Heeling inward causes immersion of the inside sidehull's inner (vertical) surface, causing a force shown in dashed-line on the left side. Combined with the reduced contribution from the inclined planing surface on the right side, the net effect is a resultant that is lower than the unheeled case. As illustrated, it is enough lower that it now passes below the CG, producing a moment which tries to return the SES to the un-heeled state. So far so good.

The problem occurs when the craft heels outward, as shown in the lower right illustration. In this case wetting of the wall-sided portion above the chine again causes the planing force resultant to angle downward compared to the initial condition. This again results in an outward roll moment, which now becomes an exacerbating moment tending to worsen the craft's attitude.

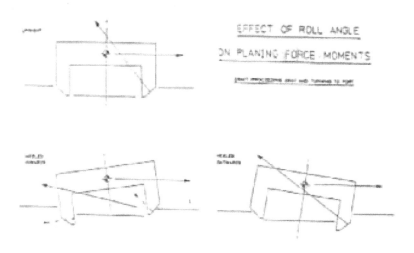

Figure 13.11: The effect that roll angle has upon the moment induced by the planing force resultant

This effect obviously depends upon the VCG as well as upon the sidehull shape. Blyth presents an interesting figure in Figure 13.12 showing the effect of VCG upon the planing force roll moment. The interesting case is the lowermost curve, corresponding to the highest KG. Note that there are two zero crossings in this case, at "A" and "C". In the region between B and C the slope is negative, and the effect of roll is to cause more roll. In fact, if the craft is in equilibrium at "B" and is then perturbed outward, it

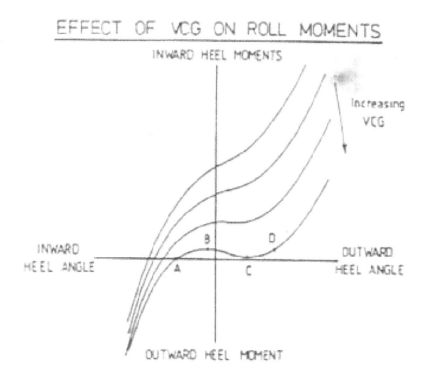

Figure 13.12: The effect of VCG on Roll Moments

will roll past "C" to a new equilibrium at "D".

Blyth took this analysis further and published some qualitative guidance showing the effect that some hull form parameters have upon the critical KG of an SES. These results are reproduced in Figure 13.13 and are generally self explanatory. Recall that in this context a higher critical vcg corresponds to a more stable ship. The following hull form changes will all increase the stability, increasing the allowable maximum VCG:

- Reducing the ratio of cushion height (depth) to cushion beam

- Reducing the inertial roll gyradius of the ship

- Reducing the sidehull width in proportion to the overall beam of the ship

279

- Avoiding deadrise above the chine, on either the inboard or outboard side of the hull

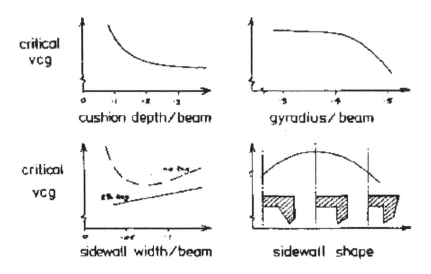

Figure 13.13: Effect of Hull Form on Critical KG

Finally, note the ways that these various effects will interact and compound each other: If the SES has the lower S-shaped stability curve, then it is possible to excite it into a roll oscillation that could mount, getting worse and worse, until eventually the craft either capsizes or at least vents the cushion.

This again is a case where the AMV demands a program of model testing to validate static-equivalent stability calculations.

13.3.3 SES Beam Sea Capsize

The final SES-peculiar issue is the possibility of beam sea capsize. Figure 13.14 depicts a time sequence of a beam sea capsize event. As shown there are two points in the sequence ("4" and "6") where capsize may occur. The physics of capsize is hard, and the only good prediction method is to model test. Mr. Lewthwaite did give me one unpublished curve which may offer some guidance on the selection of forms that do not capsize. This curve is reproduced in Figure 13.15. A fruitful research program could be pursued to validate or repudiate this curve.

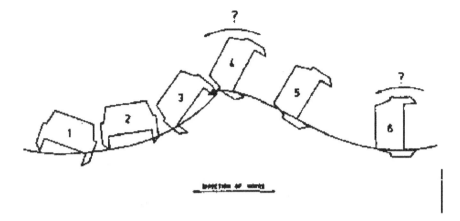

Figure 13.14: Typical SES capsize sequence in Beam Seas

13.4 AMV Stability Criteria

Stability Criteria is one of those areas where the AMV Naval Architect is navigating without a map. I have already made reference to one set of design guidance that is un-validated. This is somewhat like saying "Well I don't know, but a native told me that it might be possible to cross the river if you go about three miles upstream."

In similar spirit, I will in this section touch upon - but only touch upon - the criteria for stability of AMVs. It is, of course, absolutely vital that the practicing naval architect early find out what rules he will be required to comply with, and study those rules well to understand their implications. I hope that the paragraphs that follow will illustrate the types of implications that rules can have.

Rulemakers are human. This means not only that they can make mistakes, but also that they are approachable. It is simply wrong to wash your hands of responsibility and say "but I designed it to the rules." It is very much the innovator's burden that he must think about those rules, analyze them, critique them, improve them, and then comply with the best result of that process.

With that warning, here are some examples.

13.4.1 Intact Stability

In the domain of intact stability I will provide some design rules of thumb for SES, and then a discussion of the challenges of applying USCG rules to AMVs.

SES Rules of Thumb

John Lewthwaite and Andy Blyth have both communicated rules of thumb for SES stability. These are:

- Planing Force Resultant to cross the ship centerplane above the CG

- Lewthwaite: GM > 0.25 Cushion Beam

- Blyth:GM > 0.10 Beam OverAll

- Beam Sea Stability (Lewthwaite): KG to be below a limiting curve shown in Figure 13.15. Note that the ordinate of the curve is the ratio "Mean Sidehull Width" over "Overall Beam".

USCG Requirements

Intact stability criteria for a real AMV design project will be published by the flag state (e.g. USCG), by the Owner (e.g. USN), by IMO, by the Classification Society, or by all of them. In many cases the criteria will have been developed based on monohull practice, and will not necessarily correctly address the peculiarities of your AMV. Early dialog with the review authorities is vital.

And to reiterate my earlier statement, even when you don't disagree with the rules, simply complying with them is not ethically sufficient. You must satisfy yourself that they are indeed appropriate and adequate for your project.

With that said, let's consider some of the challenges that can be encountered in trying to comply with 46 CFR for an AMV. But again please note: The examples which follow are not intended to be a complete presentation of the USCG Stability rules! They are intended to illustrate the type of implication that the rule may have for your ship. When it comes to rules: Never work from memory - Always look it up.

USCG Criteria and Assumptions

USCG Rules (46 CFR) set requirements for initial GM, area under the righting arm curve intact, angle of equilibrium after damage, and so forth. The rules include some limits and assumptions - derived from monohull practice - that may have implications for AMVs.

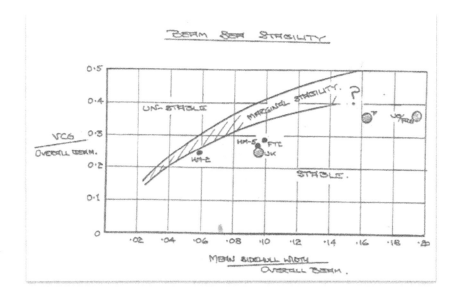

Figure 13.15: Lewthwaite's 1986 guidance on form parameters to avoid capsize. The black spots were tested craft. The large grey spots
were designs that were then under evaluation. The validity
of this curve has not been proven.

For example, consider Figure 13.16. This figure illustrates a part of the
rule that says that if maximum righting arm occurs at an angle of heel
less than 35 degrees, then you do not get to take advantage of that, and
you must "plateau" your righting arm curve to whatever value occurs at 35
degrees.

But remember the shape of a catamaran's righting arm curve: It is very
steep at the origin, and may peak well below 35 degrees. In fact, it may
peak at some low angle like 15 degrees and then diminish thereafter. Do the
USCG really mean you to not be able to take account of the tremendous
amount of righting energy represented by that peak?

Or the assumption - see Figure 13.17 - that most righting arm curves
are positive to 90 degrees, and if yours is not then you must include the
negative area and reduce your righting energy accordingly. If this is married
with throwing away a bunch of the righting energy that occurs below 35
degrees, it is easy to imagine a catamaran having great trouble complying.
Is this the Coast Guard's intent? What do other regulators say about this
matter?

GRAPH 171.055(a)

Truncation of Righting Arm Curve if Maximum Righting
Arm Occurs at an Angle of Heel Less Than 35 Degrees

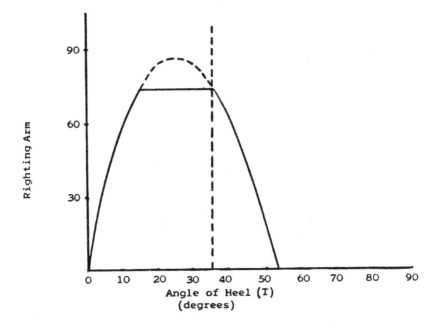

Figure 13.16: A USCG illustration based on the Assumption that Max RA occurs above 35 degrees of heel.

I shall not answer these questions. I am trying to open the reader's eyes to where there may be hidden pitfalls within so prosaic a field as intact stability analysis.

13.4.2 Damage Stability

Having used the USCG as an illustration for intact stability, permit me to use a set of USN requirements to illustrate some challenges in the domain of damage stability.

GRAPH 171.055(d)

Righting Arm Curve is not Positive to 90 Degrees and Negative Area is Included

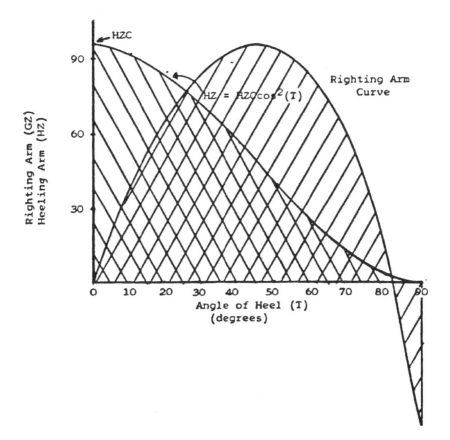

Figure 13.17: Illustrating the assumption that most righting arm curves are positive to at least 90 degrees

USN Requirements

The challenge in USN Damage Stability Criteria for AMVs lies in the required length of damage - the size of the "hole" - that the ship must survive.

USN DDS 079-1 "Surface Ship Stability" (Reference [49]) does explicitly address AMVs.

For Large SES DDS 079-1 requires a Length of Damage of 15% of the length of the ship, with an inboard penetration of 50% B - i.e. to the centerplane of the ship. This is intended to represent a fairly traditional damage case, such as being hit by another ship.

One could have interesting arguments about the 50%B component of that requirement: SES are likely to be wider than other ships of similar displacement, because of their box-like form. Does it make sense that the impaling ship will penetrate that far? Will not the low-drag form of the SES be pushed sideways before the impaling vessel reaches centerline? Or perhaps, due to the lightweight structure used on high speed ships, the ship should expect to be cut further? Perhaps a high speed ship should be designed so that it can be cut in two and both halves stay afloat? I do not practically suggest that, I merely suggest that we should think about these criteria.

But the 15%L / 50%B criterion in DDS 079-1 is not the challenging one. The challenging one is one that is intended to represent a high speed ship which strikes a glancing blow on a rock at high speed, and creates a tear in the length of the hull. DDS 079-1 requires a large SES to survive a length of damage of 50% the length of the ship - a full half! - but with a penetration of only 10% of B.

Personally, I think this is a good criterion, and ought to be applied to any high speed ship. (DDS 079-1 only applies it to SES - and if memory serves applies it even to low-speed SES!)

SWATH are not immune from tricky requirements, but in this case they almost force you to have dialog with the warrant holder: The DDS requirement for SWATH length of damage is "The same flooding length as an equivalent monohull." But what's an equivalent monohull? Is it a monohull of the same length? (That one will probably be lighter displacement than the SWATH). Is it a monohull of the same displacement? (That one will be longer than the SWATH, and thus have a longer length of damage.) Is it a monohulll with the same mission? (That would make sense - they are likely to see the same dangers, aren't they?)

DDS 079-1 doesn't mention trimarans. What is the length of damage for a high speed trimaran? Shall we take the model from the SES and require that quite-logical grazing case? Fine, then when we design a 100m trimaran we require it to survive a shallow-width but 50m-long cut. But what happens if the amas are only 25m long - corresponding to Lazausakas' guidance on what is the optimum σ value for a trimaran?

13.5 AMV Intact Stability Tests

Finally, let's consider the challenge of testing the stability of some AMVs. USCG regulations require an inclining test to verify the KG location of a new ship. Consider the 112m INCAT NATCHAN RERA.

- $LOA = 112m$

- $GMT \approx 50m$

- $\Delta \approx 3000t$

- $Beam = 30.5m$

USCG Inclining rules require a 2- to 4-degree pendulum deflection. Inclining to 2° a 3000t vessel with 50m of GM, requires 5200 tonne-meters of moment. Since the ship has a beam of ˜30m, the longest lever arm we will get for the inclining weights will be about 15m. To generate 5200 t-m of moment on an arm of 15m means that we need ˜350 tonnes of inclining weight! This is more than 10% of the total weight of the ship. It is even questionable whether any portion of the ship's deck would be strong enough to support such a concentrated weight. Clearly, a traditional inclining test on this ship is unreasonable.

Instead, we could take another logical approach: We can show that her stability is so high, that even with a ridiculous location for the KG - say, at the height of the highest deck, surely it's below that - the ship still meets all requirements. If this can be shown by calculation, do we still need an inclining?

I strongly encourage you to have this conversation early in the process, rather than having somebody read a rulebook at the eleventh hour, and impose an unreasonable and unnecessary requirement upon the ship.

AMV Stability, the bottom line:

- The physics is the same

- Cushions are de-stabilizing

- The rules are fraught with pitfalls

- The measurements are hard

14 SWBS 100 - AMV Structures

What is unique about AMV structures? It is not the material, nor the construction practices, but rather the load cases that are unlike those of conventional monohulls.

AMV loads have been studied by many experts. Rather than duplicate those researches, in this section I am going to discuss the state of the current understanding of AMV structural loads that has emerged from that work. The "vehicle" for this consensus will be the structural design rules published by the leading classification societies in the AMV field.

Once loads are known, structure can be designed to carry those loads. This process is not unique to AMVs. (The only AMV-unique aspect might be a greater emphasis on weight reduction than is found in, say, tanker design.)

14.1 Conventional Ship Load Cases

Let's once again begin by reminding ourselves of what load cases are usually used to determine the required strength of a conventional monohull.

When asked to describe a monohull's structure, the drawing that the naval architect first produces is the midship section. Why is this? It is because, at the "topmost" level, we view the ship as a slender beam subjected to longitudinal bending. We approach ship design in the belief that global loads, in particular longitudinal bending, will dominate the ship strength problem and will drive the determination of scantlings.

In the textbook case, the longitudinal bending moment is determined by balancing the ship on a static wave, such as the $1.1\sqrt{L}$ wave, and calculating a still-water bending moment due to the ship's weight distribution.

We will see that this is not the case with AMVs. It is not still-water loads that drive design, it may be local and not global loads that drive design, and the dominant loading case may not be longitudinal bending.

14.2 AMV Load Cases

In AMV structural design it tends not to be still water bending that drives design, but rather dynamically-induced loads. These dynamically-induced loads arise from speeds, speeds yield pressures, and pressures yield both global and local loads.

As our pathway through the load cases I will follow the DNV rules for High Speed Light Craft. This is not because these rules are the "best" but rather because I find their organization and details quite clear, and I think they will make a reasonable presentation of the subject matter. Other rules from ABS or other societies may be used in actual practice.

Also, this is not a presentation on the DNV rules. As I have said before: Never trust your memory or your class notes on a matter of rules - look it up. Use the most current rule book, and read all the nuances to make sure you are correctly applying the rules. This is not a lecture on DNV rules, but rather a lecture on AMV loads, using the DNV book as a roadmap.

In the practical design of advanced marine vehicles it is normal to rely on the rulebooks in the early design stages, and to then validate the rulebook loads with key measured loads from model tests. The two rulebooks I use most often are the DNV rules for High Speed Light Craft, and the ABS rules for High Speed Vessels.

An advantage of rule-book formulas is that you can quite likely code them into simple spreadsheets, and the resulting spreadsheet can be built into whole-ship design tools. Whole ship design tools can converge and balance a design rapidly, saving you from surprises later in the design process.

Our walk through the loads will distinguish between Global and Local load cases. Even here, the DNV rulebook is interesting: DNV Rules, Section 3, paragraph A100 tells us that DNV's experience is that craft under about 50m are normally driven by local loads, while larger craft are generally driven by global loads. This tells you where to focus your effort in an early-stage design.

We will begin with the global loads. There are three primary global load cases of interest to AMVs:

- Longitudinal Bending

- Transverse Bending

- Torsional Loading

14.2.1 Longitudinal Bending Modes

Longitudinal bending for an AMV is exactly analogous to that for a monohull - the craft is supported in either a hogging or sagging condition, and a bending moment is calculated.

In the case of AMVs however we treat the "support" as being not due to a static wave but rather due to a dynamic event. Figure 14.1 and Figure

14.2 reproduce DNV illustrations of these two cases. Note that they are called "landing" conditions - this suggest that DNV expects the forces to be due to dynamic events arising from operation in waves.

Indeed, the longitudinal bending mode is a dynamic situation: The craft is supported by one or two pressure patches. It is then subjected to some vertical acceleration, and a bending moment results. The global load - the bending moment - is driven by the Mass, the Acceleration, and the Lever Arm between the "supports."

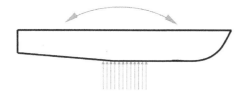

Figure 14.1: DNV "Crest Landing" condition, equivalent to hogging

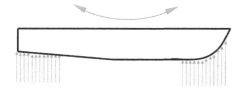

Figure 14.2: DNV "Trough Landing" condition, equivalent to sagging

The pressure patch: The "supports" are the one or two pressure patches depicted in Figure 14.1 or Figure 14.2. This patch has a "Reference Area" given by paragraph A201 as follows:

$$A_R = k\Delta\frac{(1 + 0.2\frac{a_{cg}}{g_0})}{T}(m^2) \tag{14.1}$$

Note the details of this formula: If $a_{cg} = 0$, then $A_R = k\frac{\Delta}{T}$. This means that the Reference area increases with Displacement, and decreases with Draft.

Further, as a_{cg} increases so does A_R - so clearly the next question is: What is a_{cg}?

Table 14.1: A simple parametric look at the values given by DNV's formula
for Design Vertical Acceleration

L	100	100	30	30
V	25	50	25	50
$\frac{V}{\sqrt{L}}$	2.5	5	4.56	9.13
$L^{0.76}$	33.1	33.1	13.3	13.3
$\frac{a_{cg}}{f_g g_0}$	0.24	0.48	1.10	2.20

14.2.2 The Design Vertical Acceleration

DNV's a_{cg} is a "Design Vertical Acceleration" for the craft. It represents
the dynamic load factor that the craft generates due to operation in waves.
It is defined in DNV's Section 2, page 9, Paragraph B201:

$$a_{cg} = \frac{V}{\sqrt{L}} \frac{3.2}{L^{0.76}} f_g g_0 (m/s^2) \qquad (14.2)$$

L is a reference length in meters, V is the speed in knots. "f_g" is a
factor that depends upon the Service Restriction Notation (more about
that below.)

This formula says that the design vertical acceleration increases with
Froude number, and decreases slowly (the 0.76 power) with ship length.
Ignoring the Service Restriction Factor, let's look at some values for this.
Table 14.1 shows values of a_{cg}, for $f_g = 1$, in g's, for two different craft sizes
each operating at two different speeds. The following points are salient:

- Doubling the speed doubles the acceleration

- Tripling the length lowers the acceleration by about 2:1. (Note that it
  also lowers the Froude number for a given dimensional speed, which
  further lowers the acceleration by another about 2:1.)

- The accelerations are generally in the neighborhood of one "g." They
  are neither tenths of g's nor tens of g's.

The next parameter is the acceleration factor "f_g" which is picked according to Service Restriction Notation per the table reproduced in Figure
14.3. Service Restriction notation R0 is the least restrictive, and implies
open ocean service. Restriction R5 is for lake and inland service in sheltered
water. Coastal services are usually R2 or R3.

As may be seen, the less restrictive the notation, the higher a factor must be applied to the design acceleration. Also note that the Patrol vessel has the highest factor, corresponding to the need for a warship to be driven hard even in tough conditions. It is also interesting to note that the cargo vessel has a substantially higher factor than the passenger vessel: Cargo doesn't complain until much later than the passengers do.

Multiplying Table 14.1 and Figure 14.3, we see that a 100m 50 knot passenger ferry will have a design acceleration (a_{cg}) of about half a g, whereas a 30 meter 50 knot patrol vessel might have a design acceleration as high as 15 gs. What does this mean? DNV tells us clearly: "The design vertical acceleration is an extreme value with a 1% probability of being exceeded, in the worst intended condition of operation."

As we have already seen, this value is used to determine the Reference Area for the global longitudinal bending cases. But it is also used many other places in the structural calculation. As given, the a_{cg} is the acceleration at the center of gravity. At other locations along the length of the ship the acceleration will be different. At any given station along the length, the local vertical acceleration a_v is given by $a_v = k a_{cg}$, where "k" is taken from Figure 14.4. Thus at the bow, our hypothetical Patrol boat might have to survive a load of 30 gs - which is huge.

Table B1 Acceleration factor f_g						
Type and service nota-tion	Service area restriction notation					
	R0	R1	R2	R3	R4	R5-R6
Passenger	1)	1	1	1	1	0.5
Car ferry	1)	1	1	1	1	0.5
Cargo	4	3	2	1	1	0.5
Patrol	7	5	3	1	1	0.5
Yacht	1	1	1	1	1	0.5
1) Service area restriction R0 is not available for class notations **Passenger** and **Car Ferry**.						

Figure 14.3: The selection of Acceleration Factor as a function of Service Restriction Notation and Ship Type

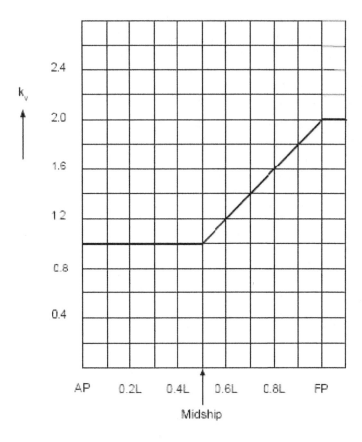

Figure 14.4: Longitudinal distribution factor for design vertical acceleration

14.2.3 Wave Height Limits

The loads don't become as high in practice as the previous paragraph suggests, because the designer is allowed to modify the speed as a function of the wave height. In my Patrol vessel example I have that unfortunate little ship maintaining 50 knots in open ocean conditions. In reality she will only make 50 knots in calm water, and she will have some drastically reduced speed in high sea states. The designer can stipulate a speed / wave height relationship in order to keep the vertical acceleration within reasonable bounds. These wave height limits become binding, and it is this that gives rise to the wave height limit table discussed in Chapter 12.

In DNV's words: *"The allowable speed corresponding to the design ver-*

tical acceleration may be estimated from the formulas for the relationship between instantaneous values of a_{cg}, *V and Hs given in 204 and 205.* ... *Relationships between allowable speed and significant wave height will be stated in the 'Appendix to Classification Certificate.'"*

Let us turn our attention to the speed / wave height relationship. DNV's formula "204" is shown below:

$$a_{cg} = \frac{k_h g_0}{1650}(\frac{H_S}{B_{WL2}} + 0.084)(50 - \beta_{cg})(\frac{V}{\sqrt{L}})^2 \frac{LB_{WL2}^2}{\Delta}(\frac{m}{s^2}) \qquad (14.3)$$

Consider for example the hypothetical case of a catamaran having the following dimensions:

- Length: $L = 105.6m$

- Beam: $B = 11.6m$

- Displacement: $\Delta = 3000t$

- $kh = 1$

- Deadrise: $\beta = 20$ degrees

Figure 14.6 shows a speed / waveheight relationship that has been constructed specifically to yield a design vertical acceleration of 0.5 g at all cases. Figure 14.7 shows how a practical limiting wave height table might be made out of that data.

V	knots		40	30	20	10	5	0
L	m	105.6						
BWL	m	11.6						
DISPL	tonnes	3000						
Type:		Cat						
kh		1						
Deadrise	degrees	20						
HS	m		3.47	4.70	5.75	7.39	8.62	10.35
V/SQRT(L)>3?			OK	NO!	NO!	NO!	NO!	NO!
acg V>3	g		0.500	0.359	0.189	0.059	0.017	0.000
acg V<3	g		0.436	0.500	0.500	0.500	0.500	0.500
acg	g		0.500	0.500	0.500	0.500	0.500	0.500

Figure 14.5: The spreadsheet used to calculate Figure 14.6

There are other important design flexibilities inherent in this method. For example, the designer could come up with a different V/H_S curve at

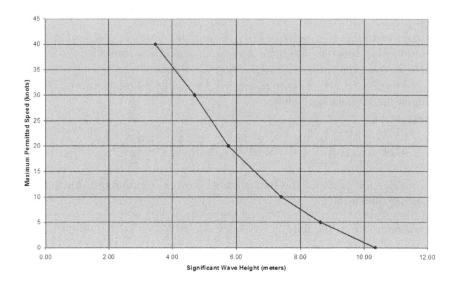

Figure 14.6: A speed / wave height relationship selected to yield constant design acceleration

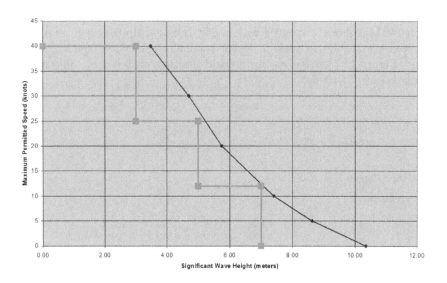

Figure 14.7: A practical limiting wave height curve overlaid on Figure 14.6

different displacements. Thus the ship might have an overload displacement, but we want the bending moment to stay within a design value. The Bending Moment $BM = \Delta a_{cg} leverarm$. Since the lever arm will be substantially the same with changes in displacement, this means that we can keep the bending moment the same at higher displacement simply by determining a new limit for a_{cg} and revisiting equation B204 to get the corresponding H_S/V limits.

Finally, for completeness' sake note that equation B204 is for speed to length ratios greater than 3. For speed length ratios less than 3, the equation is:

$acg(g's) = 6 * H_S/L * (0.85 + 0.35 * V(knots)/\sqrt{L})$

14.2.4 Design Pressures / Local Loads

The accelerations discussed thus far yield pressures, in addition to the global loads that we are going to return to. These pressures are:

- Slamming pressure on bottom

- Forebody side and bow impact pressure

- Slamming pressure on cross structures

- Sea pressure on bottom, side, & superstructure

Slamming pressure on bottom

Slamming pressure on the bottom of the ship is given by DNV paragraph C202: It applies to everything below the chine, not just below the waterline. The formula is as follows:

$$P_{sl} = 1.3k_l(\frac{\Delta}{nA})^{0.3}T_o^{0.7}\frac{50 - \beta_x}{50 - \beta_{cg}}a_{cg}(kN/m^2) \qquad (14.4)$$

As may be seen, this function says that the pressure:

- Is a linear function of a_{cg}

- Increases with Δ

- Increases with $Draft^{0.7}$

- Decreases with local deadrise angle

There is also a distribution diagram - reproduced in Figure 14.8 - that says that the slamming pressure on the bottom gets smaller toward the stern.

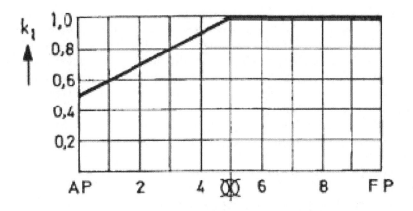

Figure 14.8: The decrease of slamming pressure toward the stern

Wet Deck Slam Pressures

The wet deck slam pressure is given in DNV C400 as:

$$P_{sl} = 2.6k_t(\frac{\Delta}{A})^{0.3}a_{cg}(1 - \frac{H_C}{H_L})(kN/m^2) \qquad (14.5)$$

This one increases from 0.5 amidships to a higher value toward the bow, according to the relationship in Figure 14.9. This figure is particularly interesting in the distinctions it draws between different vehicle types. Catamarans have a distribution factor of 1. SES and ACV get lower wet deck slam pressures, presumably because the cushion softens the impacts to some extent. Hydrofoils have the lowest pressure, which in fact is constant over the length. This is presumably because the foil control system will fly the boat to avoid slams. SWATHs have the highest pressure. This is presumably because the SWATH has very little restoring force (in the form of waterplane area) to mitigate the downward velocities that lead to wet-deck slams.

Sea Pressure

Finally, DNV's "sea pressure", which is almost "all other pressures." From equation C500 the sea pressure is NOT a function of a_{cg}, and it gets higher forward.

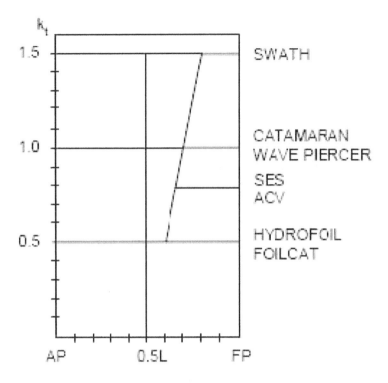

Figure 14.9: Figure 199 - Longitudinal variation of wet deck slam pressure

14.2.5 Global Loads

We started this section by trying to find something equivalent to a mono-hull's midship section. We found that the longitudinal bending moment was driven by vertical acceleration. The vertical acceleration calculation also gave us all our pressures and secondary (local) loading cases. We now return, equipped with this vertical acceleration, to the study of our three global load cases:

- Longitudinal Bending

- Transverse Bending

- Torsional Bending

— for load point below design waterline:

$$p = 10h_0 + \left(k_s - 1.5\frac{h_0}{T}\right)C_W \quad (kN/m^2)$$

— for load point above design waterline:

$$p = a\,k_s(C_W - 0.67\,h_0) \quad (kN/m^2)$$

Figure 14.10: DNV's formula for Sea Pressure

Longitudinal Bending

The hogging bending mode is what DNV calls a "crest landing" case, representing the ship coming down with the design vertical acceleration, and being supported by a single contact patch amidships. The longitudinal bending moment created by this is quite straightforward, and given by mass times acceleration times lever arm, as follows:

$$M_B = \frac{\Delta}{2}(g_0 + a_{cg})(e_w - \frac{l_s}{4})(kNm) \tag{14.6}$$

The lever arm in this case is given by the last term: $(e_w - \frac{l_s}{4})$

This term represents two half-ships: A fore half body, with some LCG_{FH}, and an aft half body, with some LCG_{AH}. Then e_w is one half the distance between LCG_{FH} and LCG_{AH}, and $l_s/4$ is the length of the Reference Area. Sketch this and you will see that it is the lever arm between the reference area and the half-masses of the ship.

The sagging mode is called, in DNV parlance, the "Hollow Landing Case." This refers to the ship landing with the design vertical acceleration on two crests, one at the bow and one at the stern, with a hollow amidships into which the ship falls. It was illustrated in Figure 14.2. The bending moment in this case is:

$M_B = \frac{\Delta}{2}(g_0 + a_{cg})(e_T - e_w)$

Where e_T is the mean distance from the center of the $A_R/2$ end areas the vessel's LCG, in meters.

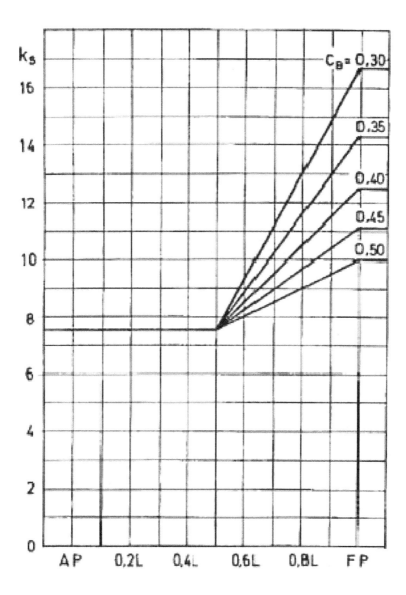

Figure 14.11: Sea Pressure longitudinal distribution factor k_s, a function of block coefficient

Transverse Bending

The second major global load case for a multihull AMV is transverse bending. This case is illustrated in Figure 14.12, and it is quite intuitive once it

301

has been shown to you. Note that in this case we are concerned not only with the bending moment, but also the shear force at the centerplane, as the two hulls try to go their separate ways.

The rulebooks give clear discussions of how to calculate these moments and forces, so I shall content myself with illustrating this load case and directing the reader to the rules for further calculational details.

The bending moment is, in the simplest case ($V/\sqrt{L}>3$, $L<50$m), given by:

$$M_S = \frac{\Delta a_{cg} b}{s} (kNm)$$

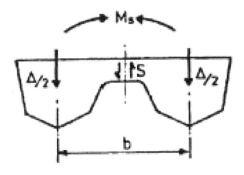

Figure 14.12: Transverse bending moments and shear force

Torsional Bending

The torsional bending mode is unique to the multihulls and may be a little harder to understand intuitively. Imagine a catamaran operating in oblique seas, where the waves arrive from 45 degrees off the bow. Now let's imagine that the wave lengths and the ship speeds are just right, so that one bow hits a crest while the other bow hits a trough. One bow is trying to pitch up while the other bow is trying to pitch down.

Of course, the two bows aren't free to take opposite paths - they are rigidly connected to each other, hence a Pitch Connecting Moment.

The pitch connecting moment manifests itself as two moments, one about the pitch axis and one transverse. This decomposition is illustrated in Figure 14.13. The formulae for the two moments are:

$$M_t = \frac{\Delta a_{cg} b}{4} (kNm)$$
$$M_p = \frac{\Delta a_{cg} L}{8} (kNm)$$

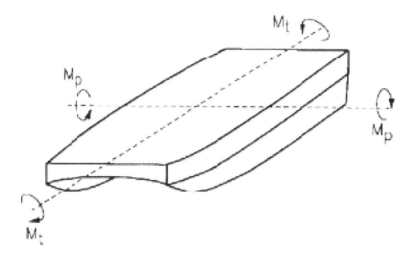

Figure 14.13: The pitch connecting moment, decomposed into Mp and Mt

14.3 AMV Load Cases Summary

Certainly there are many more rules in the DNV rule book, but this is a lecture on the AMV load cases, not on the rules.

What we have seen is the absence of a still-water wave, but instead we find that loads are driven by dynamic events (acceleration) not static still-water bending moments. Hull girder bending (i.e. midship section) may still be the defining consideration, but the accelerations are high enough that they make pressures that are high enough that for smaller ships it is local loads that dominate. The conventional-ship "midship section" case (longitudinal bending) is in our world replaced by three bending cases:

- Longitudinal

- Transverse

- Torsional

15 SWBS 119 - Design of Air Cushion Skirts

This is another section where an entire textbook in its own right is needed. In this section I shall be able to do little more than acquaint the reader with the issues involved in the design of a skirt system, and educate the student on the basics of inflatable structures.

An ACV skirt system is a complex engineered structure. Figure 15.1 illustrates the components of one such system, and I reproduce it here merely to underscore the complexity of the system.

Very early air cushion vehicle (ACV) concept development began with Sir Christopher Cockrell's first ACV in the late 1950's which had no flexible seals to contain the air cushion which supports the weight of the vessel. Thus there was a very low daylight gap between the ACV's hard structure and the land or water below it. This makes it difficult for the vehicle to clear obstacles or navigate in rough terrain or waves. The flexible skirt concept soon followed, enabling the ACV to keep its very low daylight gap to minimize the cushion power required and yet permit the vessel's hard structure to be above nominal perturbations in the water or land surface.

It was later in the 1960s that the bag and finger skirt appeared for the ACV. The bag was needed to provide a more reactive surface on the seals, enabling it to better react to waves and other surfaces while the ACV is pitching or rolling. All operational ACV's now use bag and finger seal or skirt systems.

An amphibious ACV must use air propulsion since it is amphibious and can not have water propulsion. This is unfortunate since it can not take advantage of the factor of 800 in fluid density between air and water. (There was an ACV ferry which used water propulsion with a skeg and water prop, but it was limited to in-water operations only.) In the early 1960's, in consideration of this and other factors, Mr. Allen Ford patented his Captured Air Bubble Vehicle while working at the Naval Air Development Center in Warminster, PA (US Patent 3146752.) The CAB vehicle had movable seals at the bow and stern and rigid sidewalls on each side to better contain the air cushion and to support water propulsion. A 10-ton, 50 ft. long test craft was designed and built in the shipyard in Philadelphia. This XR-1 testcraft had seals that were comprised of two athwart-ship panels

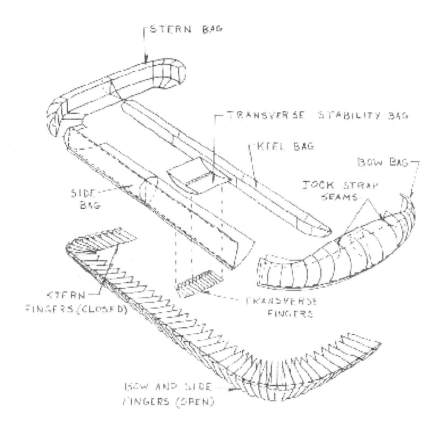

Figure 15.1: A hovercraft skirt system, simply to illustrate the complexity of this engineered product

that were held down by the pressure in the CAB's cushion and their lower movement was limited by cables or down stops. The XR-1 and its 50% increased beam modification, known as the XR-1A,became the very early workhorse for testing out new CAB (later to be called a Surface Effect Ship) seal systems, dynamic controls for its lift system to improve motions and new waterjet systems.

Most of the seal development on the XR-1C, -1D and -1E were planing type seals but with much improved flexibility and dynamics relative to the earlier rigid planing seals. This type of seal was eventually designed, built and installed on the SES-100A1 testcraft and was the original seal design for the Rohr 3KSES.

But local structural loads at the fasteners of the down stop cables and the fiberglass planing sections became difficult to solve in the time-driven program, so the final design changed to a bag and finger bow seal, similar to the seal on the SES-100B testcraft and other lesser-size SES vessels being built at that time. Again, the fingers were to help in containing the cushion air while operating in seas and the bag provided pitch-up forces and moments in addition to the pitch moments provided by the increased cushion length due to the slope of the fingers, for the craft in higher seas.

15.1 Purpose and Types of Skirts

As with so much of the AMV design, we begin with the teleology of a skirt: What is the purpose of the skirt on a powered-lift craft? First, let it be well understood that skirts are not necessary. Sir Christopher Cokerel's first hovercraft did not have a skirt. But the skirt was quickly invented and retrofit to the SR.N-1.

During the early development of the SES, we learned about the physical phenomena that impact the properties that we want in bow and stern seals. The seals contribute to craft performance in many ways:

- They must be adequately flexible to respond to various shaped waves to minimize airflow leakage, in various headings to the sea;

- They must be able to conform to the slope of the cushion pressure-generated wave which varies with Froude Number. This has a strong effect on low speed resistance.;

- An SES gets much of its pitch restoring moments from the bow seal, plus the bow seal must be able to restore its shape to its original condition after a pitch excursion so that the lift from the cushion is restored rapidly;

- To minimize the frictional resistance of the sidewalls of an SES, the seals optimally are configured such that their lowest point is at or very near the keel of the sidewalls at the bow and stern. This changes with speed however, since the cushion generates a wave pattern inside the cushion and the slope of this pressure generated wave varies with Froude Number.

- The cushion generated wave which approaches the stern seal is very 3-dimensional and its shape varies with Froude Number, requiring a degree of athwartships flexibility.

- The seals must not generate suction forces. The stern seal is much more prone to this issue than the bow seal;

- As the seals move vertically due to wave action, they must be able to survive the structural snatch-loads which are felt as the seal rapidly moves down after the wave passes;

- The lift system mechanism which strongly assists in driving the seals back to their lower position after they have been moved by the waves, must force the seals down very rapidly;

- The seals must be durable relative to material wear and abrasion;

- The seals' concept and design must be able to be manufactured and repaired, preferably at sea.

Simplifying this, the ideal skirt system will:

- Retain the air bubble - reduce air leakage

- Have no drag

- Conform to waves without exciting ship motion - weightless / massless

- Assist with pitch stability

To accomplish the first of these - Retain the air bubble - the skirt must:

- Resist cushion pressure

- Retain desired geometry

- Be impermeable

This would be well accomplished by steel, for example. But now consider the goal of being dragless. To accomplish this we want a perfect geometry of water contact, so that there is no wetted surface of skirt. In a static hovering condition this might be possible with rigid skirts, but the shape of the interface surface varies with speed, so we need a seal that also changes shape with speed. And there are waves: In waves we want the skirt to deflect out of the way instantaneously. This requires the skirt to be inertialess, or massless - not an attribute of steel.

Finally, we want the skirt to assist in providing pitch stability for the craft. This means that the bow skirt will have a forward slope to it, so that when the bow pitches down there is some forward shift in the center of pressure, resulting in a pitch restoring moment.

To address these multiple goals, many types of skirt have been invented and tried. My list includes No skirt, Air Curtains, Water Curtain, Pericells, Fingers, Bag and fingers, Stayed bags, and Transversely stiffened membranes. But despite this broad range, these can be collapsed into three major types, which I shall address in turn:

- "Virtual" Skirts

- Rigid Skirts

- Inflatable Skirts

15.1.1 Virtual Skirts

I class both "Peripheral Jets" and "Water Curtains" as "Virtual skirts" because there is no physical structure retaining the cushion, instead it is retained by an inertial barrier formed by a mass of fluid - either air or water.

Peripheral jets

A peripheral jet system consists of a thin slot around the perimeter of the craft, and a high pressure jet of air blowing through this slot toward the ground. The momentum of the air jet is sufficient to retain a positive pressure inside the perimeter, in the cushion area of the craft.

The governing relationships, that give us the required flow and pressure from the jet, are given by Yun & Bliault [8] in the page that I reproduce as Figure 15.2. Note that the peripheral jet also supplies the air to the cushion - there is no separate lift fan system in addition to the jet fans.

A homework assignment can be given in which the student will use these relations to find the lift power for a small number of hypothetical hovercraft. As will be seen, the problem with the peripheral jet method is that it requires a lot of power: The air jet must be given enough momentum to retain the cushion, which requires a substantial jet pressure and flow rate.

Water Curtain

The Water Curtain concept is similar to that of the peripheral jet. But whereas the peripheral jet combines both cushion retention and cushion creation into a single air flow stream, the water curtain does require a separate air cushion fan system. It then uses the water curtain only to retain the cushion. The idea of the water curtain is to use a mass of falling water to produce the pressure barrier that retains the cushion. The innovation is that water will have no drag when it touches the ocean, because it will

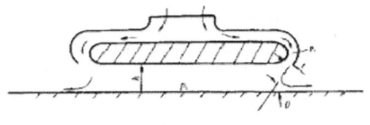

$$hp_, = p_, V_,^2 / t \, (1 + \cos \theta) = p_, V_,^2 x / h \qquad (2.1)$$

where $p_,$ is the cushion pressure (N/m²), $p_,$ the air density (Na²/m³), t the width of nozzle (m), h the air clearance (m), $V_,$ the mean velocity of jets (m/s), θ the angle between the centre-line of the nozzle and the craft baseline (°) and

$$x = (1 + \cos \theta) t / h$$

Then the total pressure of the jet can be expressed as

$$P_, = 0.5 p_, V_,^2 + f p_, \qquad (2.2)$$

$$Q = V_, t L_, \qquad (2.3)$$

Table 2.1 Coefficient f relative to h/t

h/t	f
1	0.75
2	0.65
3	0.54
~4	0.50

Figure 15.2: Yun & Bliault [8] description of the governing equations for peripheral jets.

"disappear" into the ocean. It will also conform perfectly to waves. The problem of course is that to create the water curtain we must lift seawater up from the surface and then eject it downwards with enough momentum to seal the cushion. It turns out that the energy required to do this is much larger than the savings due to elimination of seal drag. There are no water-curtain craft in existence that I know of.

15.1.2 Rigid Skirts

The peripheral jet and water curtain ideas are two ideas that seem to get re-invented once each generation. Rigid skirts are another similar case. Many people have thought of using a rigid structure to retain the air cushion, and then articulating that structure on a system of hinges and springs to give it the desired dynamic performance.

Obviously a rigid skirt is a good solution to the permeability goal, and it requires no power for the skirt itself. Early rigid skirts consisted of simple hinged plywood panels fitted at the bow and stern of an SES. The first generation of this simply hinged the panel at the top with a door hinge. The problem is that the cushion pressure acting behind this panel results in a large force, and simply makes the panel into a plow, eliminating the resistance advantages of the SES.

To solve this, the inventors switched to a "balanced" type design, where the panel was hinged about a mid-chord, and not at the edge. This results in a panel with good conformance to the 2D surface.

The problem now is that any athwartships shape to the wave is met with a single monolithic panel, which is then plowed through the wave. So the clever inventors conceived of segmenting the panel athwartships into a system of several rigid fingers, that look something like piano keys. These must of course be of balanced design, as well. As the individual keys move, they must have some means between them so that air doesn't escape between adjacent fingers. This requires some kind of side panels to close the gap, and these side panels will rub on each other. The friction thus introduced will reduce the conformability of the fingers, reducing their effectiveness and increasing their drag.

In practice, nobody has yet overcome these solutions with a system that is superior to the inflatable fabric skirt.

15.1.3 Inflatable Fabric Skirts

I classify fabric skirts into six basic families, each of which will be discussed below. There are:

- Curtain

- Transversely Stiffened Membrane

- Bag

- Pericell / Jupe

- Finger

- Bag and Finger

Curtain Skirt

A flexible skirt helps reduce the air flow required to support the craft. Making this skirt of fabric will help reduce the weight of the skirt and may reduce its tendency to plow and drag, because of its flexibility. The simplest type of flexible fabric skirt would be a simple curtain hanging down from the wet deck. This skirt would have to be tensioned at the bottom in order to hold it down or it will simply blow up under the influence of the cushion pressure. Some of the hold-down effect can be attained by making the skirt go around the full perimeter of the craft and making it somewhat conical - tapering downward. The sloped sides of the cone and the inherent geometry of the cone will help to keep the skirt in place. Unfortunately, the same forces that keep a curtain skirt in place also stiffen the skirt and make it more likely to drag by plowing.

Transversely Stiffened Membrane

Imagine an SES curtain skirt that is about the size of the door on a two-car garage. Imagine that it is secured along the entire top edge, and also at the two sides, but not at the bottom. Now imagine it subjected to 1 psi of pressure on one side. Obviously it is going to bulge outward, and will no longer be a simple 2D shape. To alleviate this bulging some practitioners have experimented with transversely stiffened curtain skirts (see US Patent 4333413 [50].) In this case long thin flexible battens are included in the skirt, spanning the full width from side to side. These battens help reduce the transverse bending of the fabric. They may also be tethered to the ship structure for further geometry control. Very few TSM skirts have been built, and little is known about the potential of this system.

Bag Skirt

At some sort of "opposite extreme" from a curtain would be to surround the full perimeter with an inflatable "horse collar" or "inner tube" all the way around. This skirt will work. There will be a tradeoff between pitch stability and plowing / drag - a higher pressure inflation will make it stiffer, yield more pitch stiffness, but also result in higher drag. Indeed at the limit - infinite inflation pressure - this becomes simply an analog of a rigid non-compliant skirt.

In actual practice skirt inflation pressures are far below infinity, but they must still be somewhat higher than the cushion pressure. Bag skirt systems are common (indeed, ubiquitous) as stern seals in SES. In this application they are usually inflated to 5% - 15% above the cushion pressure. This produces a very soft bag which is easily deflected by incident waves.

Pericell / Jupe

The next type of skirt consists of the use of a series of smaller conical structures. Recall that the curtain concept evolved into a single all-around cone. The Jupe concept is to use instead a chain of smaller cones. These are called "jupes" which is simply the French word for "skirt." Each pericell or jupe looks something like the garment called a "tulip skirt." A series of these jupes surrounds the cushion - sometimes in combination with a common bag section, as illustrated in Figure 15.4.

Figure 15.3: One type of inflated skirt.

The pericell yields good vertical stiffness if the cells are conical in shape. The drawback to a pericell is that the portion of the hem of the skirt that is concave-forward is shaped to scoop water when in motion, which can cause drag, skirt damage, or other undersirable behavior. This can be mitigated by slanting the tips of the cones somewhat so that the forward facing edge is slightly higher than the aft-facing edge.

Finger

Somewhere between a curtain and a pericell lies the concept of the finger skirt. A fabric "finger" is a half-cylinder of fabric, suspended from the wetdeck at an angle of about 45 degrees from the vertical. The half-cylinder has its convex face outward, concave toward the cushion pressure. The

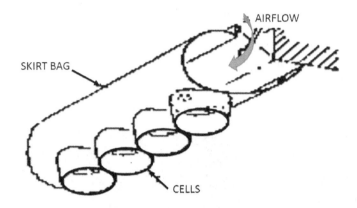

Figure 15.4: A Pericell and Bag (or Jupe and Bag) skirt system

finger skirt may be considered to be a derivative case of the curtain skirt, where a single large curtain is replaced by a series of multiple curtains. This philosophy is illustrated in Figure 15.5. One may also imagine a finger skirt as consisting of only the outboard halves of a series of pericells.

Bag and Finger

In Figure 15.4 I illustrated a bag-and-pericell skirt system. I have also said that a finger skirt may be considered equivalent to half a pericell. In that case it is unsurprising to introduce the bag-and-finger combination, illustrated in Figure 15.6.

15.2 Basics of Inflatable Structures

I have tried to explain the evolution or teleology of various skirt concepts, but to further understand the skirts on Powered-Lift AMVs we must learn a few basics about inflatable structures. There are two simple facts that are all-important:

- The force balance on a uniform membrane will always result in a circle (or segment)

- The stress in an inflated segment is directly proportional to the radius

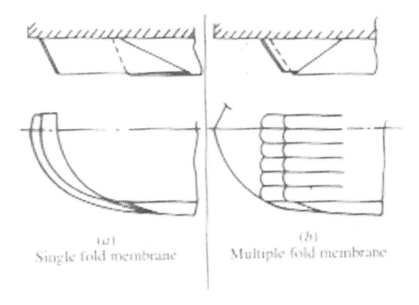

(a)
Single fold membrane

(b)
Multiple fold membrane

Figure 15.5: The finger skirt (right) explained as a derivative case of a single curtain skirt.

These two facts can be clearly seen if you imagine a fabric having zero stiffness. If it is subjected to a uniform load like a cantilever beam it will of course deflect. With zero stiffness the resultant forces at the endpoints can only be in the direction of the fabric - in pure tension. Consider the case shown in Figure 15.7. Here the diameter of the circle is equal to the space between the supports. The total force acting on the restraints must be the integral of the pressure over the girth of the bag, resolved into X and Y components. It is easy to see that the Y components will cancel out due to symmetry, and the X component will be equal to $P * D$. PD is thus the total reaction, which is the sum of the two endpoint force $R1$ and $R2$. Thus $R1 = R2 = P * Radius$. Now what is the tension in the fabric? Is it not simply the reaction force R? Thus the tension in the fabric is $t = R1 = R2 = P * Radius$.

15.3 Basic Design of SES Skirts

The most common SES skirt system today consists of full-depth fingers forward, and a multi-lobed bag aft. Since this system is common, and

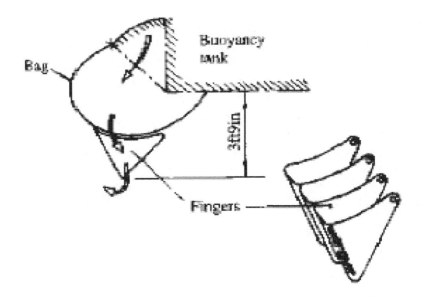

Figure 15.6: A bag-and-finger skirt system

since an AMV-acquainted naval architect might be called upon to develop a concept design fairly quickly without recourse to consultants, I shall provide an overview of how to design this type of skirt system.

15.3.1 SES Bow Finger Skirts

Beginning with the bow skirt system: This system consists of fingers extending the full depth (or height) from the wet deck to the designed cushion depression. The key features of such a system are:

- Semi-cylindrical fingers

- Angled to the waterline

- Restrained at tips

The fingers are half cylinders facing convex-forward. The diameter of these cylinders is determined by the strength of the skirt fabric, since as we have seen, the tension in the fabric will depend simply upon the cushion pressure times the finger radius. As a design parameter, SES in the 40-80m size range have cushion pressures of 0.75 - 1.5 meters of water, and have finger diameters of about 1 meter. Thus for a "starting point" we may take a design value of hoop stress from these values, and scale to any particular project's cushion pressure to estimate that project's finger diameter, assuming the same hoop stress as the design value.

The fingers do not descend vertically - they form some angle with the vertical. This angle is a major source of the pitch stability of an SES, and is also important to the drag of the skirts. The common design angle is 45 degrees. I have seen and tested angles from 30 to 60 degrees, and there is nothing to recommend them at this time. A flatter, more horizontal, skirt angle will increase the pitch stiffness because it yields more shift of the center of pressure, but it will likely increase the wetting of the skirt and thus skirt friction. It probably reduces skirt wavemaking drag because it forms a more gentle entry angle for the cushion, viewed in profile. A more vertical angle reduces pitch stiffness, but may also increase drag because it presents a more blunt entrance angle to the cushion pressure.

Imagine if the fingers were simple half-cylinders, attached at the wet deck, and angled 45 degrees from the vertical. Now subject them to cushion pressure on the aft face. Clearly they will buckle and fold forward unless they are restrained in some manner. The restraint must hold the finger tip aft against the force of the cushion pressure, and it must not yield any effective force acting up the long vertical axis of the cylinder, as this would simply cause the finger to crumple vertically in buckling. Thus we see that the restraint could be as simple as a pair of ropes attached to the lower aft corners of the skirt and lead to the craft structure, provided that these ropes form an angle of at least 90 degrees with the finger axis.

In practice, ropes are not used for this purpose because of the next required feature: The fingers must seal against their neighbors. If the fingers were simple half-cylinders, then for any non-zero deflection they would open a gap between themselves and their neighboring finger, and cushion air would leak out of this gap. Therefore the half-cylinder of finger has straight-line extensions aftward. These flat panels of fabric bear against the neighboring fingers (or the craft sidewalls) even when the finger has moved appreciably.

In practice, the fabric extensions are carried all the way aft to serve as the restraint "ropes" mentioned previously. Of course, this fabric exists at every point along the edge of the finger, not merely at the tips, so the restraint force is applied distributed over the length of the finger, which avoids load concentrations. The resulting finger geometry is depicted in Figure 15.8.

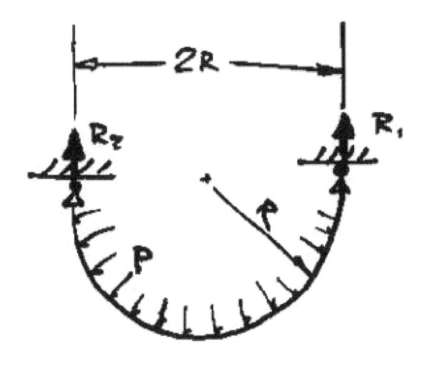

Figure 15.7: Basics of inflatable structures

15.3.2 SES Stern Bag Skirts

An SES stern bag is a simpler geometry. A two-lobed bag is shown in Figure 15.9. The stern bag geometry is dominated by the ratios of pressures inside and outside the chambers of the bags. Consider Figure 15.10, which shows a simple single-lobe configuration. The key to this geometry is that the tension in the fabric must be the same at every point along the perimeter - there is no mechanism for increasing or decreasing tension except at the end points. The controlling points then, which are under the control of the naval architect, are the two endpoints and the point of tangency with the sea surface (labeled "t".) I have simplified this geometry such that the aft endpoint "A" is vertically above point "t". This makes the aft

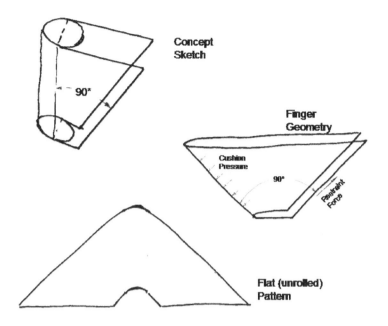

Figure 15.8: Drawings of generic SES bow-finger geometry

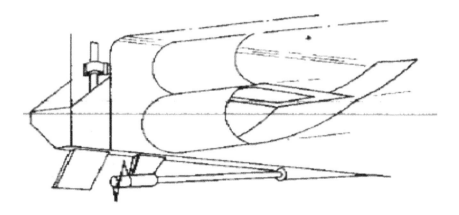

Figure 15.9: A two-lobed SES bag-type stern seal

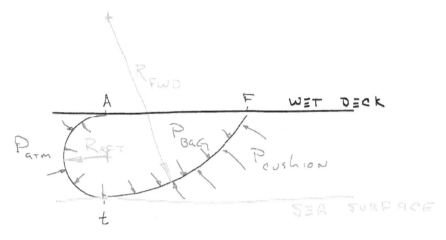

Figure 15.10: Definition sketch for a simplified case of the geometric balance of a stern bag seal

lobe of the bag a complete semi-circle. The student can generalize this geometry to other cases, and indeed Larry Doctors has provided a complete generalization in his work, see earlier in this text.

There are two different pressures acting on the bag in Figure 15.10. The aft face of the bag sees a pressure which is P_{aft} = Bag Pressure minus Atmospheric Pressure. This is higher than the pressure seen on the front curve of the bag, which is $P_{forward}$ = Bag Pressure minus Cushion Pressure.

Now, since the tension at point "t" must be the same on the forward and aft parts of this point, this means that P_{aft} times $Radius_{aft}$ must equal $P_{forward}$ times $Radius_{forward}$.

With the height of the cushion known, and the pressures known, the designer can then find the resulting two radii, and from these can calculate the amount of fabric (girth) required to make the stern seal. Typical stern bag pressures are 5% - 20% above cushion pressure, with the philosophy being "the lower the pressure the better."

This geometry is sufficient to design a practical bag skirt - by applying these relationships one establishes the required girth (arc length) of the lobes of the bag, and thus the skirt. There is one additional parameter to be considered, however: When the bag is washed aft by a wave, it may be carried aft of the transom of the sidehull. When this occurs the ends of the bag will be exposed to atmosphere and the bag will deflate. This in turn leads to loss of cushion pressure, etc.

The only sure way to prevent this is to ensure that the cushion walls extend far enough aft that the bag will still be contained by these walls, even when it is flattened against the wet deck. In terms of the figure below, this would mean that the transom should be located at a point "x" aft of "A", such that the distance $|xA| + |xF| = $ arc-length(At) + arc-length(Ft).

Note that a bag-type stern seal does not have end caps - the edges of the fabric simply slide along the rigid craft structure. This means that, especially when un-inflated, the stern bag can fill with sea water. To drain this water a series of small holes are included, lying along the line of tangency "t". To prevent the holes from catching the water at high speed, and thus tearing the fabric bag, a simple flap of cloth on the outside (attached forward and loose at the aft end) covers them. This flap of cloth is called a "feather" - despite not looking like one at all!

It is appropriate here to comment on the hydrodynamics of the stern bag. In an ideal stern bag there is a small daylight gap between the stern bag and the sea surface at point "t". This gap is an interesting mess of stuff to analyze. First, the inflation of the seal wants to press the bottom edge against the water surface while the air in the cushion wants to lift the bottom edge and get out. In addition, there is a venturi effect as the cushion air jets through the gap, and this causes a suction that pulls the bag down to the surface.

An interesting flutter can be created: when the venturi pulls the bag down to the surface it closes the gap. When the gap closes, the flow stops, and the venturi suction goes away. Absent this suction the static forces reassert themselves and the bag pulls up from the surface an inch. This of course causes cushion air to flow once more, recreating the venturi, and pulling the bag back down to close the gap. In practice, the result of this is a resonant mode that is exactly like a "whoopee cushion" or clarinet reed.

Balancing those two forces plus the venturi effect is indeed the *mètier* of a seal design specialist. In addition to the seal pressure and cushion pressure affecting the shape of the inflated seal, the venturi under its gap has to be tweaked with the correct approach angle and number of drain holes that bleed air in to the venturi and, sometimes, even a trip or step along that edge to reduce the suction or down force created by the venturi. This seal-water interface pressure distribution and magnitude has been measured in tests and correlated with a rather complex model of the stern seal dynamics. FYI...the pressure at the gap can drop below atmospheric pressure.

The issue in these investigations was not seal drag, but instead we wanted to understand most or all of the mechanics affecting steal stability; some bounced and some didn't and unstable ones created serious ride quality problems. However, it stands to reason that the ones that were unstable contributed a drag component since each bounce resulted in contact with the water surface.

In calm water, a correctly designed lobe seal probably has little if any drag contribution since it is not contacting the water surface and has no perceptible effect (visual anyway) on surface elevation forward of, or under, the sea itself.

SES Specialist Rick Loheed (private communication) provided some interesting comments which seem to fit nowhere else, so I include them here for the reader's benefit:

"Our tuning trials were typically for optimizing motion controls, but during our testing we always did a "Stern Seal Delta-P Vs Speed" sensitivity test without any other variables changing. When venting the cushion for control in waves it was found running slightly tighter than design allowed a higher vent valve "effective bias" dynamically, yielding a little more bidirectional control and keeping more cushion pressure longer as the seas got bigger, resulting in higher speeds with the same power.

"In observing the seals during initial trials, I used to adjust the Delta P until it was observed to contact the water, and then back off just enough so it didn't too quickly arrive at an operating point near the optimum. I was seldom wrong by much."

Mr. Bill McFann (private communication) added: *"I always made it a point to make the yard install windows into the cushion for seal observation. I also thought they should be there so the crew could check the seals for damage, but the classification societies typically made them remove them. I think maybe a couple put covers over them and managed to keep them. Others argue for a video camera. It is never as good and they can fail- I still think the crew needs to be able to observe the seals directly.*

"In calm water, watching the stern seal gap is fascinating because the water smoothly flows beneath it, then rapidly begins the rise to the surface because the pressure is off and the venturi is helping turn the flow. Everything is a bright green- it is seldom as dark in there as you would think. It looks like the smooth inside curl of a breaking wave just under and behind the aft lobe. This of course means the venturi "wraps" around the lower lobe also- it does not exit flat as if shedding from a transom.

"Typically I could not see any spray from inside the cushion- the surface was usually very smooth. It may have gone more turbulent near the surface where atomization events like ligaments turning into droplets could occur more readily as the escaping air tears at it."

15.4 Skirt Forces

The forces in a skirt system are in three classes:

- Internal forces

- Attachment Forces

- Dynamic Forces

15.4.1 Internal forces

We have already seen that the basic internal force in a fabric structure is due to the inflation pressure. If there are no stays, wires, restraints, etc., then the structure will take on a circular shape and the fabric will be loaded to a Hoop Stress which is equal to Stress = Radius X Pressure / Thickness. This stress drives the selection of the number of fingers or lobes in a seal, and is driven by the allowable stress of the skirt fabrics.

15.4.2 Attachment forces

Attaching a fabric skirt to a rigid craft is not simple. The challenge is to try to create an attachment system that is continuous, e.g. a bolt rope, in order to avoid stress concentrations. The purpose of this attachment is to provide the restraint forces needed to hold the skirt in place against the "thrust" caused by the pressures. Figure 15.11 through 15.16 illustrate a few details used in accomplishing this.

Sometimes the skirt must include point-load restraints, such as stays or webbing straps. In this case the manner of attaching these items requires doubler sections and grommets. Indeed, skirt manufacture is very like sail-making, and most of the techniques for handling reinforcements in a sail are also used with skirts.

Skirts are not made in single elements. Especially in the case of fingers, it is desirable to make the skirt in segments. In the case of fingers it is normal to have each finger fitted with a removable "cuff" at the bottom edge. This is where most of the finger wear takes place, and with this method one can simply remove and replace the cuff, rather than the whole finger.

Similarly, bag segments in a multi-lobe stern seal may be made removable. This makes possible afloat maintenance, as seen in Figure 15.18. Segmented construction also results in controlling the weight of any single component, easing maintenance and installation.

Attaching segments to each other is usually accomplished using point loads. I have seen both bolted attachments and lacings used equally successfully. Bolting is straightforward, and requires local reinforcement. In the case of lacings the system consists of simple grommets through which a cord is woven and tied, exactly like tying a shoelace or corset.

15.4.3 Dynamic forces

There are static forces due to inflation. There are local issues due to attachments. There are also some very important dynamic forces present in even a simple skirt system. The most important dynamic force is flagellation.

Flagellation takes place especially at finger tips or any other similar unsupported edge. The trailing edge in such a situation will flutter and flap, exactly like a flag in a breeze. The trailing edge itself flaps back and forth several times a second, subjecting itself to high accelerations. Finger tip accelerations have been measured to exceed 8000 g's - Wow! This gives rise to a form of fingertip wear that looks exactly like abrasion - see . It also, however, gives rise to internal heat build up that can destroy the skirt fabric from the inside. The rapid flexing of the fabric results in energy that shows up as heat, and can cause burning or melting of the fibers or of the rubber coatings of the fabric. This problem gets worse as fabric gets thicker, because the thicker fabric has a harder time shedding this internal heat.

15.5 Skirt Failures

Skirts do fail. Most failures are simply wear, rather than catastrophic-event type failures. Wrinkling, delamination, and abrasion of finger tips is common and should be provided for by designing removable cuffs. Bow skirt wear rates are on the order of one millimeter of fabric lost per hour of high speed (¿40 knots) operation. This yields finger cuff replacement intervals of about 1000 underway hours.

Stern bag wear occurs at the edge of the tube. Many designers use a wear drape or feather in this location. Wear can also occur along the feather used to cover the drain holes. Stern seal wear rates are much lower, with stern seal repair / replacement intervals on the order of 5000 ship hours.

It is possible to tear a seal, say by striking a log or other obstacle. If tearing is expected to be a problem due to the nature of the operation then it is recommended to design skirts that include rip stops (similar to crack arrestors in early steel shipbuilding.)

In the extreme case, a skirt can blow out. Blow out is usually associated with snatching or snap-back loads in a wave encounter. This results in a force which is basically the same as the restraint forces and steady state pressures TIMES a dynamic load factor.

15.6 Skirt Materials

The materials used in modern full-size skirts are virtually the same as used in inflatable boats: Natural rubber reinforced with nylon, etc. The issues in selecting a skirt material are:

- Strength, to withstand the skirt forces (including local loads)

- Heat tolerance, to withstand the heat generated by flagellation

- Flexibility, to yield the comforming behavios sought in a skirt

- Adhesion between the fiber and the matrix, to ensure long fabric life

- Repairability, including the feasibility of using adhesive patches, stitching, etc.

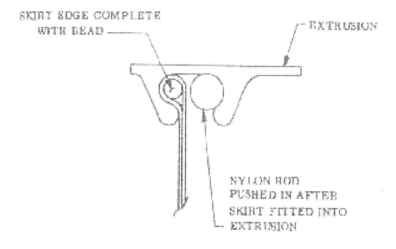

Figure 15.11: A bolt-rope type skirt attachment scheme

Figures 15.19 and 15.20 present some data on skirt materials produced in China, and skirt life from a sample of existing craft, taken from Yun & Bliault [8]. In the western world the only skirt maker I know of is Avon Engineered Fabrications, a division of Avon Rubber (the makers of the successful Avon line of inflatable dinghies.)

At model scale, some model makers use sailcloth to fabric skirts in the towing tank. There is debate as to whether this is satisfactory, as sailcolth

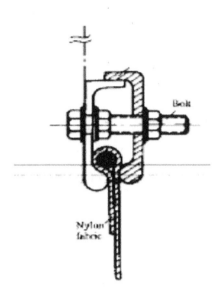

Figure 15.12: A bolt-rope scheme involving a bolted clamping system

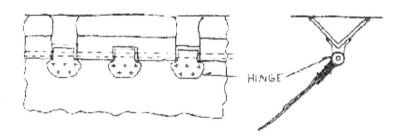

Figure 15.13: Mechanical hinge attachment concepts

will not have the same weight / stress / strain properties as scaled full-scale fabric.

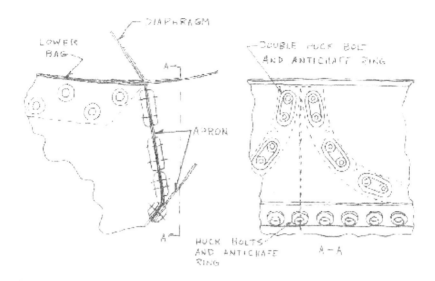

Figure 15.14: Bolted attachment of skirt segments, with bolts protected by anti-chafe rings (see next figure.)

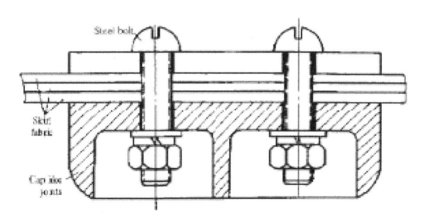

Figure 15.15: An anti-chafe collar

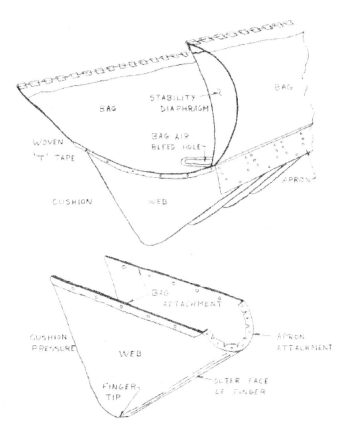

Figure 15.16: Attaching the fingers to the bag

Figure 15.17: An SES bow skirt, where the wear at the tips of the fingers due to flagellation is clearly visible

Figure 15.18: Showing the afloat detachment of two bag segments from a three-lobed stern seal

Skirt fabric designation	Units	648-1	57703
Width of coated fabric	mm	810	830-840
Thickness of coated fabric	mm	2	2.5
Specific weight of coated fabric	kg/m2	2.19	2.57
Peel strength - Original	N/15 cm)	660	980
Peel strength - 1 week's soak in fresh water	N/15 cm)	160	
Peel strength - 20 days' soak in fresh water	N/15 cm)	160	920
Peel strength - 1 week's soak in 10% salt water	N/15 cm)	350	
Peel strength - 20 days' soak in 10% salt water	N/15 cm)	260	
Breaking strength of coated fabric - warp	N/15 cm)	7140	4920
Breaking strength of coated fabric - weft	N/15 cm)	6270	6200
Tearing strength of coated fabric - warp	N	770	1490
Tearing strength of coated fabric - weft	N	930	1300
Application		Small and medium size ACV or SES	Medium-size ACV and SES

Figure 15.19: Data table from Yun & Bliault describing two skirt fabrics available in China [8]

Craft name	Craft weight (t)	Maximum craft speed (knots)	Cushion pressure (Pa)	Skirt height (m)	Coated fabric (kg/m)	Tension strength (N/cm)	Tear strength (N)	Skirt life (hours)	Notes
SR.N4 h	200	70	2521	2.4	2.9-4.6			5000+	
Mk.2 f					4.5	8722	1875	100-400	
V1.1 b	110	46	2992	1.68	2.4	5690	893	5000+	
f					1.36			300-1200	
VT.2 b	106	60	2900	1.68	2.44	5690	863	9000+	
f					1.36			300-1000	
SR.N6 b	10.8	54	1256	1.22	1.36				
Mk.1 f					3.0	5690	893	200-750	
HM.216 h	20	35		1.0	3.0			8000+	
f					1.2		935	300-1500	
BH.110 b	138	35						700	
7202	2.8	24	981	0.5	1.5	2943	432	300	5802U fabric
731-II	5.0	52	1170	0.75	2.1	5886	383	250	6408 fabric
716	45.0	50	1471	1.6	2.1	5886	483		6108 fabric
722-1	65.0	50	2451	1.6	2.6	4905	1177		5791E fabric

b = big, f = finger

Figure 15.20: Data table from Yun & Bliault describing skirt materials and life from some built SES and ACV [8]

16 SWBS 200 - Propulsors

The propulsion of AMVs is not different from the propulsion of any other marine vehicle, except perhaps for the speed of interest. This course will therefore focus on introducing propulsors common for high speed craft, and their selection and installation. The propulsion question is straightforward - efficiently generate thrust - so the discriminator questions tend to be "How do you steer?" and "How do you reverse?" I will begin by introducing two important types of screw propeller for high speed craft, and will then discuss waterjets. The physics of propellers and waterjets will not be much discussed, because this subject is well treated elsewhere in the naval architecture curriculum.

16.1 The Propulsion Task - Required Thrust

The task of the propulsor is, obviously, to generate a thrust equal to the resistance of the ship. This requires us to know the resistance of the ship, and this was covered earlier. But it is not enough to simply take that resistance and pass it to the propulsor designer as his task - there are a couple of nuances that must be accounted for.

16.1.1 Resistance Margin

First, let use remember all the uncertainties in resistance that we touched upon in Chapter 9. The resistance of the ship is not perfectly known, and the wise designer adds a margin to his resistance estimate to ensure that his ship does in fact attain the contracted speed. These margins vary with individual practice, but they are generally about 15% before model tests, and 8% after model tests.

Some commercial practitioners don't formulate their margin in that way but prefer to take it as a speed margin. In this case the practice I have seen is to take the resistance by reading the curve one knot too high - that is to say, for a 40 knot design case, design for the resistance estimated at 41 knots.

Whichever approach is used, the result is to take the basic resistance estimate and translate it into an estimate for use in propulsor design.

16.2 Thrust Required

Having generated a resistance estimate to be used in propulsor design, how much propulsor thrust shall we require? There are a couple more margins to be accounted for on this side of the task.

16.2.1 Hump Thrust Margin

Some high speed craft have a pronounced hump in the resistance curve, at a speed much lower than the design speed - around a length Froude number of 0.30. This hump is particularly troubling because it is possible to be stuck on the low side of it, resulting in a maximum speed less than half the craft's potential speed. (I was involved in one case of a 40-knot catamaran who had experienced enough weight growth that her hump drag had risen and she could no longer get over hump. This 40-knot ship would labor along at full throttle at just under 15 knots.)

Unlike top speed, we need to clear the hump with some substantial extra thrust. This is because we want to accelerate through the hump - we do not want to operate there in steady state. This requires that the thrust at hump must exceed the drag at hump by some margin.

Hump thrust margins that I have seen are of two styles: One group imposes a percentage type, requiring that the thrust at hump speed must be at least 25% greater than the resistance at this speed. Another approach is to require a certain acceleration rate at hump, such as 0.05g. This is equivalent to saying that the thrust must be greater than the resistance at hump speed by an amount equal to $1/20^{th}$ the craft's weight. This latter formulation is interesting: Consider the case where the craft Lift to Drag (L/D) ratio is 10:1 at hump speed. In this case the estimated hump drag is equal to 1/10 the craft weight. An acceleration margin of 0.05g would add 1/20 the craft weight to this, in other words a hump thrust margin of 50%.

16.2.2 Thrust Deduction

The next component is the thrust deduction. For propellers this is just the same as with conventional hulls and therefore is covered in other textbooks. The thrust deduction is a factor that - for a conventional propeller in the behind condition - states that the attained thrust is usually a few percent lower than obtained in open water. It may be argued that key here is "the behind condition"', and that some propellers such as surface-piercing propellers may operate in such clean flow that they are in nearly open-water conditions, and thus experience a thrust deduction of zero. This is, of course, application specific.

For waterjets the thrust deduction is usually negative - meaning that a waterjet generates slightly *more* thrust when installed than when in an open-water condition. This is sometimes attributed to the interaction of the waterjet inlet flow upon the hull, arguing that it is not properly a thrust effect (a negative thrust deduction) but is actually a resistance effect (a drag reduction.)

16.3 Propulsor types

Up to this point we have developed a resistance estimate, and we have now translated that into a curve of required thrust. Now it's time to pick a propulsor.

16.3.1 Propellers

I shall address two types of "unconventional" propellers, the fully submerged cavitating propeller, and the surface piercing (or partially submerged) type. My treatment will once again be "practical" and not theory-based. I encourage the further study of the theory of these propellers, but I hope to provide the student with a working knowledge which he can bring to those theoretical classes and thus gain even more from them.

Fully-submerged cavitating propellers

First, let us remind ourselves of what cavitation is. Cavitation is the limiting value of pressure on the suction side of the blade on any propeller. When the pressure on the blade drops below the cavity pressure (approximately the vapor pressure of water) the water changes state and becomes a gas (steam.) Cavitation is not necessarily bad. *Unsteady* cavitation is undesirable for many reasons, mostly because the collapse of the cavitation bubble causes erosion of the metal propeller blade and vibration-exciting impulses. But *steady* cavitation is an acceptable operating regime for a purpose-designed propeller.

What defines the likelihood of cavitation? In order to determine if cavitation will take place, we calculate a cavitation number for the propeller. A low number means cavitation is likely. The cavitation number measures the pressure on the suction side of the foil. For a propeller, at the radius $= 0$, the cavitation number is given by:

$$\sigma = \frac{p_a + \rho g h - p_v}{\frac{1}{2}\rho(V_A)^2} \tag{16.1}$$

Where:

p_a = atmospheric pressure

p_v = water vapor pressure at the temperature of interest

g = gravity

h = submergence of propeller shaft

V_A = local velocity

Consider the formula for cavitation number: The only role of ship speed is in the denominator, squared. What this is meant to illustrate is simply that cavitation number falls - and thus cavitation becomes more likely - as ship speed increases.

Technically the velocity in the denominator of the cavitation number formula is the inflow velocity at the blade, and not simply the ship speed. It is instead a vector sum, a triangle with one side being formed by the ship speed and the other side by the rotational (tangential) velocity of the blade. The hypotenuse of this triangle is the sum of the squares of these two speeds, and thus the velocity at any arbitrary radius is:

$$\sigma = \frac{p_a + \rho g h - p_v}{\frac{1}{2}\rho[U^2 + (2\pi n r)^2]} \tag{16.2}$$

Where:

n =revolutions per second

r =radius

U =ship speed

Substituting the propeller speed parameter "J" into the RPM term, at say, the 0.7Radius, we obtain:

$$\sigma = \frac{p_a + \rho g h - p_v}{\frac{1}{2}\rho U^2} * \frac{1}{\left(1 + \left(\frac{\pi 0.7}{J}\right)^2\right)} \tag{16.3}$$

This shows that cavitation number will further reduce as J is reduced - e.g. as RPM or diameter are increased. Figure 16.1 shows approximate domains of cavitation as the local cavitation number, vessel speed U, and advance ratio J vary. Again, high speeds, and low J, lead to cavitation. The "breakpoint" between cavitation and no cavitation is somewhere in the neighborhood of $\sigma = 0.10$.

Newton-Rader Propellers

The Newton-Rader propeller series is a propeller specifically designed to operate in a fully-cavitated condition. If, during initial design, the naval architect has determined that a fully cavitating propeller should be investigated for his project, there is enough data in the Newton Rader series literature (and Figure 16.3) to complete the initial sizing of a Newton Rader

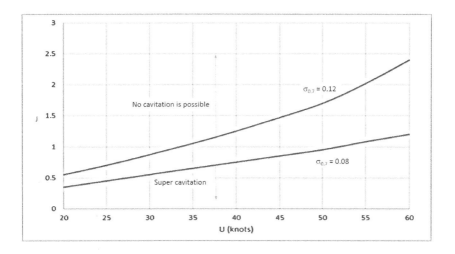

Figure 16.1:

propeller. Of course, subsequent design phases will want to use more detailed treatments of these propellers, but this is enough to get started.

Surface Piercing propellers

The other interesting class of propeller is the surface piercing propeller. These props have demonstrated very high efficiencies at high speed. Indeed, they are ubiquitous in the race boat community. Figure 16.4 illustrates a race-boat installation.

The surface piercing propeller, as the name implies, is not fully submerged. Figure 16.5 is an illustration of a test stand, but it also serves as a good definition sketch of an SP installation. The key parameters (in addition to the normal factors of blade number, shape, etc.) are the percent of propeller immersion and the shaft rake. The immersion is usually expressed as a percent of propeller diameter, such as "50% immersed" (which would mean immersed right up to the shaft centerline.) When a surface piercing propeller blade enters the water, the suction side of the blade is in a solid bubble of air. This is the bounding value for lowest possible suction-side pressure: p-atmo. Figure 16.6 shows the air cavity wake behind an SPP operating at about 35% submergence.

Rose & Kruppa, in 1991 [51] & 1993 [52], published design data for a systematic series of surface piercing propellers. Data was presented for shaft angles of 4, 8, and 12 degrees, and for P/D ratios of 0.9 to 1.6. Figure

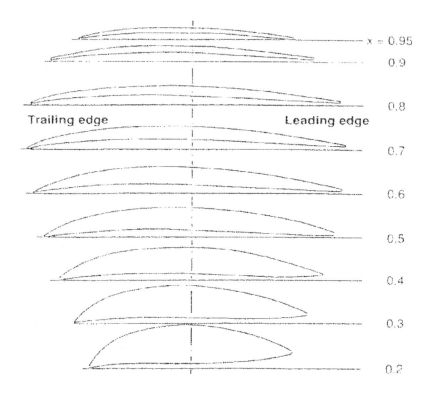

Figure 16.2: Newton-Rader series blade section shapes

16.6 is the summary of performance for the 12-degree case, with P/D = 1.75. The data given provides efficiency and KT data for four values of immersion ratio, and a wide range of J. The data is in the form of J vs KT/J^2, as this makes the plotting easier. From this data a practitioner can easily extract KQ (from KT and η), and thus solve for a design condition. Note that the full set of Rose & Kruppa data is available in the NavCAD propeller selection module, as propeller type "SP."

16.3.2 Waterjets

The best single reference I know of on waterjets is Allison, [9], which I usually distribute as a class handout. There are also excellent discussions in Faltinsen [4] as well as in the marine propulsion courses in a naval architecture curriculum. My focus in this course will be upon practical considerations of commercial waterjet units.

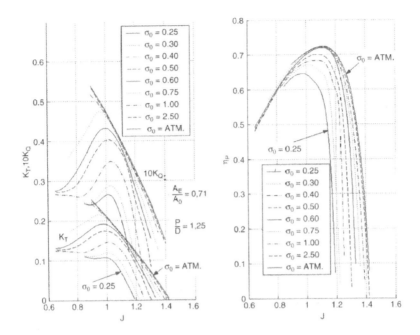

Figure 16.3: Performance characteristics of the Newton Rader series propellers. From [4]

Waterjet Hydrodynamics

To understand waterjets properly it is helpful to have a few hydrodynamic concepts solidly in mind. I shall introduce these here quickly.

The Gross Thrust of a waterjet is derived entirely from the aftward momentum of the discharged water: Gross Thrust $= T_G = \dot{m}V_J,$ Where:

- $\dot{m}$ = nozzle mass flow rate

- V_J = Jet Velocity

This is exactly like the high-school physics problem of propelling a boat by throwing rocks over the stern. In effect, the waterjet discharges a continuous stream of "rocks" (water particles) out the stern.

Unlike the rowboat full of rocks, however, the waterjet has to scoop up the discharged mass from the water as it goes by, just as if the hypothetical boy in the rowboat had to pick the rocks up from the bottom of the river as he passed them. This means that the "rocks" (water) must first be brought

Figure 16.4: Twin surface-piercing propellers on a race boat

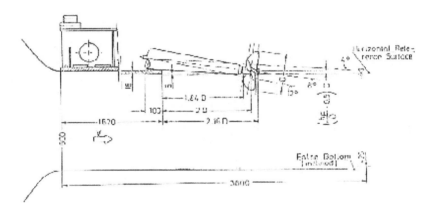

Figure 16.5: A Surface-Piercing Propeller test rig, which illustrates the major parameters of the SPP

up to the boat speed, before it is discharged at jet speed. This represents a loss in net thrust, that is easy to write down:

Figure 16.6: A photo of the air cavity behind a surface piercing propeller

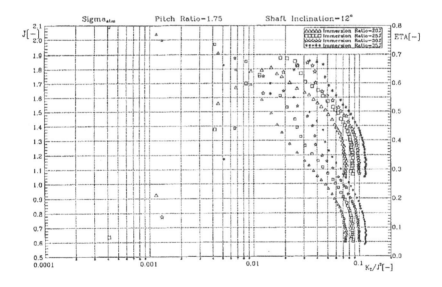

Figure 16.7: Rose & Kruppa data for a surface piercing propeller with P/D=1.75, 12° shaft angle

Net Thrust $= T_N = m(V_J - V_S)$
Where:

- $\dot{m}$ = nozzle mass flow rate

- V_J = Jet Velocity

- V_S = Ship Velocity

Waterjet Efficiency (Theory)

The equations above relate to the discharge velocity and mass rate of the jet plume. The efficiency of this process depends primarily upon the ratio of the discharge velocity to the inlet velocity: The Jet Velocity Ratio (JVR)$= V_J/V_S$. At a JVR of 1.0 the theoretical efficiency would be 1.0, but unfortunately the thrust would be zero. Practical values of JVR and jet efficiency are shown in Figure 16.8, taken from Allison.

Waterjets Pump Types

To create a waterjet of the highest attainable efficiency, the jet designer will select from an appropriate pump type. The three major classes of pump are:

- Centrifugal

- Mixed-Flow

- Axial

A centrifugal pump works by "flinging the water out" centrifugally, and then capturing that flow in a volute and directing it in the desired direction. Figure 16.11 depicts the generic case of a centrifugal pump. The key fact is to recognize that the flow arrives along the centerline axis of the pump (in the "doughnut hole" so to speak) and leaves along the perimeter of the doughnut. Compare Figure 16.11 with the photograph in Figure 16.9 and familiarize yourself with the shape of a centrifugal pump.

By contrast in an axial pump the flow does not make this abrupt change of direction, but arrives and departs along the same line, along the main axis of the pump. Again, see Figure 16.10 and compare with Figure 16.12.

Mid way between an Axial- and a Centrifugal-flow pump lies the mixed-flow pump, which combines both flow types.

Each of these types has a particular regime in which it is most efficient. A pump designer may select a pump type based on indication given in a Cordier Diagram such as reproduced in Figure 16.13. In this diagram the following terms are used:

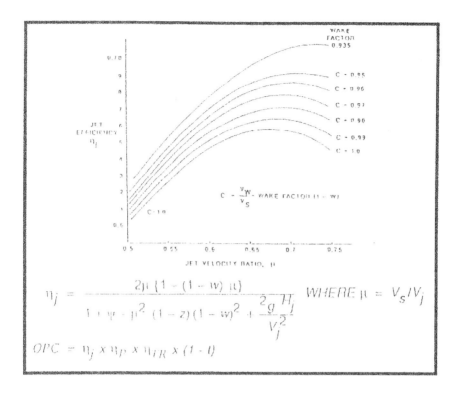

Figure 16.8: Theoretical waterjet jet efficiency, for practical values of JVR and wake fraction, from [9]

- Specific Speed: $N_s = nQ^{0.5}/(gH)^{0.75}$

- Specific Diameter: $D_S = D(gH)^{0.25}/Q^{0.5}$

- n=rps

- D=diameter (m)

- H=head (m)

- Q=flow (m^3/s)

Figure 16.14 and Figure 16.5 present two depictions of a commercial mixed-flow waterjet, with Figure 16.14 being a hydrodynamic depiction, Figure 16.15 a mechanical depiction.

341

Figure 16.9: An early waterjet based on a centrifugal-type pump

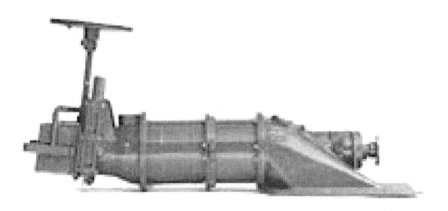

Figure 16.10: An early waterjet based on an axial-type pump

Commercial Types

I have greatest personal familiarity with three major manufacturers of commercial mixed-flow waterjet propulsors: KaMeWa / Rolls-Royce, Wartsila LIPS, and HamiltonJet. Commercial manufacturers sell waterjets from a catalog in standard sizes. The sizes are generally cataloged by waterjet

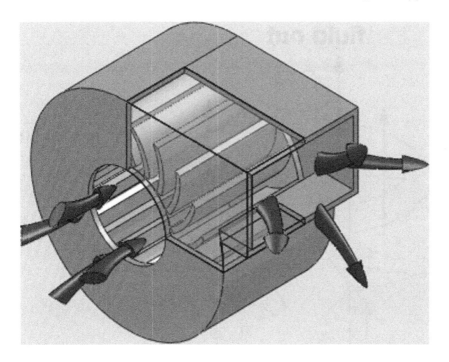

Figure 16.11: A textbook illustration of a centrifugal pump

diameter. This may be either the impeller diameter or the discharge nozzle diameter, or the inlet diameter, so be careful: One manufacturer's "100 cm" jet may be another man's "125 cm" unit.

It is not far wrong to assume that all waterjets have the same thrust loading (Thrust / D^2), so the diameter is determined by the thrust required. One can also use this rule of thumb to guesstimate the diameter that will be needed, during the very earliest days of a design project if you know the diameter and thrust (or power) from another successful installation. Thrust and speed of course yield power, so higher power jets are also larger diameter jets. Figure 16.16 reproduces a recent KaMeWa jet range for their S-series units. The dimensions of the S-series are shown in Figure 16.17. The KaMeWa model number is simply the inlet diameter in centimeters (not shown in this figure.)

Design Considerations

When installing the waterjets there are three concerns that I wish to mention. A lot of guidance is available from the jet manufacturers as the design

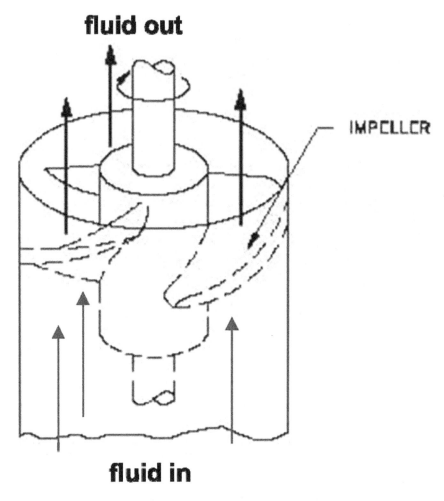

fluid out

IMPELLER

fluid in

Figure 16.12: Textbook illustration of an axial pump

progresses, but it is helpful in the earliest days if the concept design can take account of a few features to help ensure success in later stages. I consider as paramount the need to:

- Avoid inlet suction

- Avoid impeller cavitation

- Avoid inlet cavitation

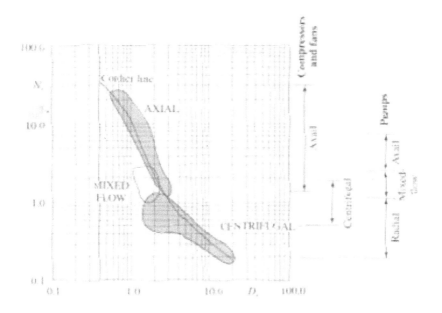

Figure 16.13: A Cordier diagram of pump regimes

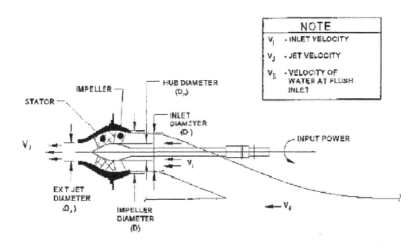

Figure 16.14: A mixed-flow waterjet

Inlet Suction - the Waterjet Capture Area A waterjet draws its inlet water from a large volume upstream, called the "capture area" of the jet.

345

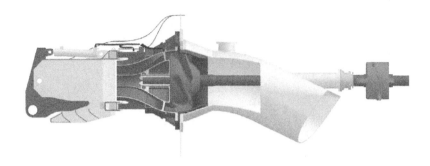

Figure 16.15: A mixed-flow waterjet

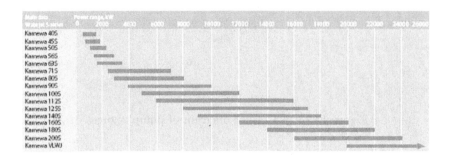

Figure 16.16: KaMeWa S-Series units, relating size (model number) to power

At rest, a simple thought experiment will make it clear that this must be a nearly circular volume. As speed increases this volume becomes narrower and more focused forward. Figure 16.18 illustrates the capture area found in one CFD simulation of waterjet performance. It is important to keep this area free from air, by ensuring that the jet is deeply submerged from the free surface, from any air cushions, and from any entrained air sheets under the hull bottom. The size of the capture area can also be important in shallow water, with many sometimes-humorous stories arising when jets are run at high throttle in shallow water, and suck a wide range of unlikely objects into their inlets.

Inlet Cavitation - Inlet Pressures The next concern is to avoid cavitation within the inlet. The jet manufacturer will design the inlet as part of his scope of supply, but it is helpful to understand what his concern is. Figure

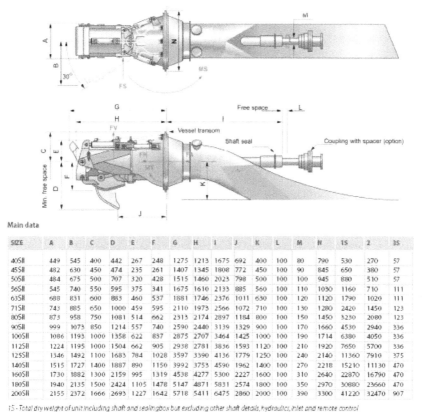

Main data

SIZE	A	B	C	D	E	F	G	H	I	J	K	L	M	N	1S	2	3S
40SII	449	545	400	442	267	248	1275	1213	1675	692	400	100	80	790	530	270	57
45SII	482	630	450	474	235	261	1407	1345	1808	772	450	100	90	845	650	380	57
50SII	484	675	500	707	320	428	1515	1460	2023	798	500	100	100	945	880	510	57
56SII	545	740	550	595	375	341	1675	1610	2133	885	560	100	110	1030	1160	710	111
63SII	688	831	600	883	460	537	1881	1746	2376	1011	630	100	120	1120	1790	1020	111
71SII	743	885	650	1000	459	595	2110	1973	2566	1072	710	100	130	1280	2420	1450	123
80SII	873	958	750	1081	514	662	2313	2174	2897	1184	800	100	150	1450	3230	2080	123
90SII	999	1073	850	1214	557	740	2590	2440	3139	1329	900	100	170	1660	4530	2940	336
100SII	1086	1193	1000	1358	622	837	2875	2707	3464	1425	1000	100	190	1714	6380	4050	336
112SII	1224	1195	1000	1504	662	905	2938	2781	3836	1593	1120	100	210	1920	7650	5700	336
125SII	1346	1492	1100	1683	784	1028	3597	3390	4136	1779	1250	100	240	2140	11360	7910	375
140SII	1515	1727	1400	1887	890	1150	3992	3753	4590	1962	1400	100	270	2218	15210	11130	470
160SII	1730	1882	1300	2159	995	1319	4538	4277	5300	2227	1600	100	310	2640	22870	16790	470
180SII	1940	2135	1500	2424	1105	1478	5147	4871	5831	2574	1800	100	350	2970	30680	23660	470
200SII	2155	2372	1666	2693	1227	1642	5718	5411	6475	2860	2000	100	390	3300	41220	32470	907

1S - Total dry weight of unit including shaft and sealingbox but excluding other shaft details, hydraulics, inlet and remote control
2 - Weight of water in pump and inlet duct
3S - Hydraulics including PTO pump &dry weight)

Figure 16.17: Geometry of the KaMeWa S-series

16.19, from Faltinsen [4], illustrates the surface pressures found on the walls of waterjet inlet. Figure 16.20 illustrates the same phenomenon using a 3D CFD representation.

Looking at the colors in Figure 16.20, we may see that there is a large region of negative pressure on the forward ramp area of the inlet, and also on the tip of the lip. These are two areas where real waterjets exhibit cavitation erosion of the inlet.

In designing an inlet to avoid cavitation on the ramp, the designer will try to make the inlet longer. This is effective, but it increases the weight of entrained water in the inlet, thus reducing craft buoyancy.

Designing to avoid lip cavitation is much harder, and becomes increas-

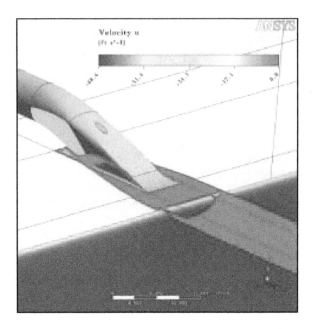

Figure 16.18:

ingly challenging as speed increases. Indeed, in the days of the 100-knot SES program the attention was focused on variable geometry lips, so that the shape could be adjusted to the proper cavitation-free design as a function of speed.

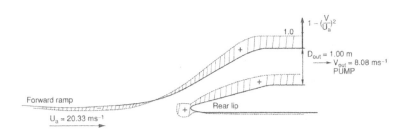

Figure 16.19: A profile of a waterjet inlet illustrating the pressures experienced on the boundary, from [4]

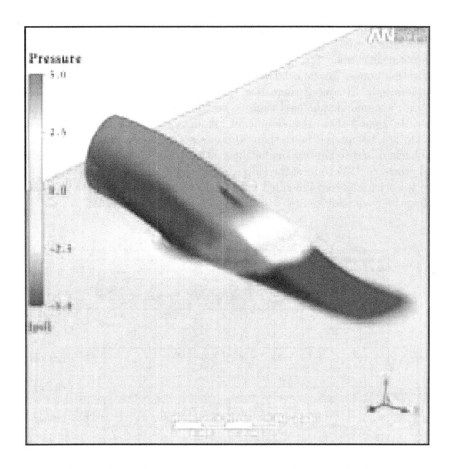

Figure 16.20: Surface pressures in a flowing waterjet inlet

Waterjet Impeller Cavitation Boundaries Finally, in addition to inlet cavitation which is addressed during inlet design, the naval architect must avoid impeller cavitation through loading of the impeller.

Figure 16.21 illustrates a commercial quote obtained from Rolls Royce for a project in 2005. (This is real data, but do not use it as design data. It is specific to the particular project quoted.) The chart has axes of speed at the bottom, and thrust along the vertical. The red lines are curves of the thrust required for the particular ship at 2890 and 3135 tonnes displacement.

The series of solid black lines, roughly horizontal, represent the thrust produced by a suite of four KaMeWa jets, at various power levels and all speeds. As can be seen, for example, at 4 x 21240 kW the 2890 tonne ship will attain a speed of about 47.5 knots. Now note the dashed curves and the notations "zone 1" "zone 2" and "zone 3". These represent impeller cavitation zones. In zone 1 there is no impeller cavitation and operation is unlimited. In zone 2 there is a small amount of cavitation, and Rolls Royce recommend that operation be restricted to less than 500 hours per year. In zone 3 there is a significant amount of cavitation, and this zone should not be entered more than 50 hours per year. As you may imagine, the background to this particular illustration was that now that the weight had grown to 3135 tonnes, the ship was too solidly into zone 2, and a larger size waterjet was recommended in order to reduce the thrust loading back below the cavitation limits

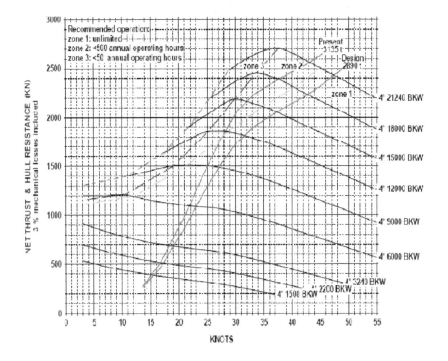

Figure 16.21: A KaMeWa quotation for a specific project, involving quadruple size 153 waterjets

Many waterjet craft do enter into "zone 2" but this is intended to be a transient condition. Figure 16.22 is Faltinsen's illustration of this, showing that one may enter zone 2 during the short time of transiting a resistance hump.

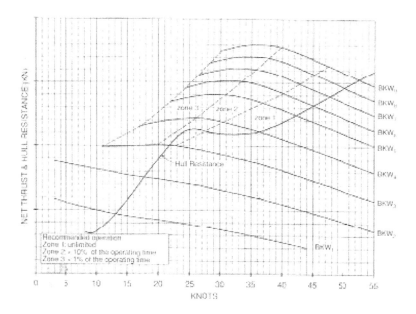

Figure 16.22: Illustrates the case of a craft entering the cavitation zone for a brief period for an event such as hump transit. From [4]

Waterjet scaling issues
Interesting recent (2011) work by Julie Lin (reference needed) highlights one more issue with waterjet application, which is the difficulty of scaling the waterjet properly in model tests. My own knowledge on this subject is gleaned from one or two of Dr. Lin's briefings and is thus superficial at best, but it is sufficient for me to present here a small overview and some cautions.

In brief, it is clear that the waterjet performance will depend on the thickness of the boundary layer that is ingested into the waterjet. This boundary layer has the effect of lowering the average inlet velocity, as well as contributing to the non-uniformity of the flow into the waterjet impeller.

However there is a further boundary layer contributed by the walls of the inlet tunnel itself.

These boundary layers will be improperly scaled in a model test, due to the improper matching of the Reynolds number, as is well known. Further,

351

the very small model waterjet impeller blades, will not be operating at full scale Reynolds numbers, nor will their ratio of Reynolds number "model / full-scale" be the same as that of the hull or the inlet. This of course introduces a scaling error in a model test self-propulsion result. One means around this might be to use a large cavitation tunnel instead of a towing tank, so that the body Reynolds numbers can be matched. The problem here is that the hull forward of the inlet will not be present (LCC tests do not usually include a full hull model) and thus the wake development into the waterjet will not be the same as the full-scale vessel. The upshot of all of this, is that the waterjet model-scale self-propulsion tests are fraught with increased scaling difficulties.

Waterjet RPM Relationship

Another interesting feature of waterjets is the thrust vs RPM relationship. Most naval architects are familiar with the model of a screw propeller as a solid screw, boring its way through the water. This model is not applicable to a waterjet. A waterjet is much better to be thought of as a constant power device, wherein the amount of power consumed depends only upon RPM, and not upon the ship's speed through the water.

This is illustrated by Faltinsen in the figure shown as Figure 16.23. This is a cleaned-up version of a KaMeWa quote document, and it very clearly shows that thrust is extremely flat with ship speed, depending almost entirely on shaft RPM.

I have seen this relationship used in a practical manner during a dispute with an engine manufacturer over whether the engine was in fact delivering the power contracted for. The waterjet serves as a power dynamometer: If the jet won't reach RPM "X", then the engine is not putting out power "Y".

Waterjet overall effectiveness

I have shown a few previous graphs of waterjet thrust versus speed. Figure 16.24 presents one such graph, wherein I have translated particular points into values of Overall Propulsive Coefficient, EHP/BHP. As may be seen, the waterjets have quite respectable overall efficiency, in the range of 60-70%.

Let me take this opportunity to reiterate some caveats: A waterjet manufacturer will choose from several impellers, several nozzles, and several inlets. Performance prediction curves are developed for each specific application, and will differ slightly from case to case. Any curves from a different

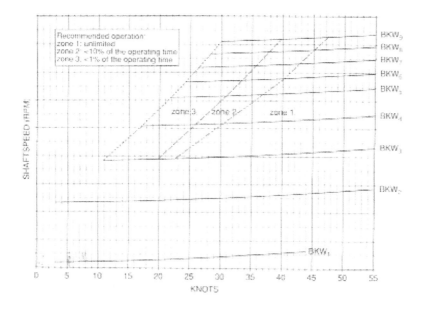

Figure 16.23: Relationship between power, RPM, and speed for a waterjet

project, or from a brochure, or from these course notes, must be taken as indicative only.

Waterjet Arrangement

The arrangement of waterjets into the hull is quite straightforward (which is one of the reasons they are so popular.) There are however a few issues that bear to be touched upon.

Waterjets side by side: Multiple jets can be installed side by side. In such a case the designer should seek consultation with the manufacturer. This is because the capture areas will interact and / or compete. Also, if on a sloped hull (deadrise) the pressure gradients at the jet inlet will differ due to the differing hydrostatic head on one side or the other. This may result in slight differences in the jet loadings from neighboring jets (probably negligible, in practice.)

Also, when installing multiple Jets, consider that not all jets need to be steerable / reversible. "Booster Units" are available from most manufacturers, consisting of waterjets without their steering and reversing components. These units are lighter than the fully steerable units, and cheaper too.

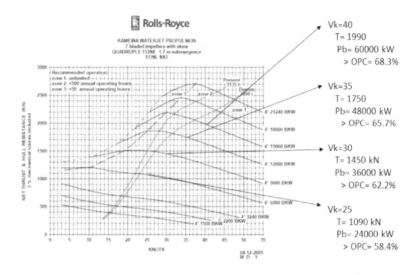

Figure 16.24: Attained waterjet performance values for one design project

Waterjet Weight

Waterjets are heavy. The waterjet unit consists of a substantial mechanical component, plus the inlet assembly. These weights are given by the jet manufacturer in their catalogs. In addition to the weight of the jet unit, the naval architect must also deal with the weight of the water entrained within the unit. (One might argue that this should be treated as a loss of buoyancy volume, but it is more common to treat it as a carried weight.) The water weight can be obtained from most manufacturers, or it can be estimated by calculating the volume of a cylinder of water having the length and diameter of the jet inlet duct.

Waterjet Structural Loads

Waterjets have unusual load paths - or at least they were unusual to their early adopters. Consider where the thrust of a waterjet is generated: Some portion of it is generated by the impeller and is transmitted down the propeller shaft in the conventional manner. But there are very substantial pressures acting on the stator blades and on the walls of the duct, that are transmitted into the ship structure.

One brand of waterjet includes the thrust bearing within the impeller hub, and then transmits the thrust out through the stator blades. For a jet of this type this means that all of the jet thrust is delivered to the ship's transom structure. Make sure your stern scantlings can take this load, and can do so while maintaining the very small deflection tolerances needed for a waterjet installation.

Figure 16.25 illustrates a Wartsila LIPS jet and shows that this unit uses a conventional thrust bearing on the shaft forward of the inlet. This reduces (but does not eliminate) the amount of force that is transmitted via the transom structure, but it also adds the necessity for including a thrust foundation for this bearing.

The waterjet steering loads are substantial as well. They will generally be located as point loads acting on the steering axis, and on the attachment points for the hydraulic actuators of the steering / reversing assembly.

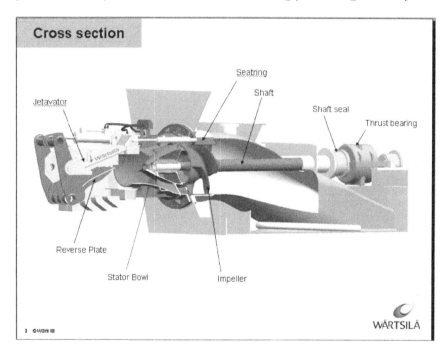

Figure 16.25: A Wartsila jet, clearly showing the location of the thrust bearing

Waterjet Scope of Supply

Another attraction of the waterjet is that they are usually sold as turn-key suites of equipment. The vendor's scope of supply normally includes:

- Waterjet Pump

- Inlet Design

- Including fabrication for Hamilton

- Design Only for Rolls Royce & Wartsila

- Control System

- Actuators

- Hydraulic Pack

- Bridge Controls

- Probably Includes Engine Controls

- FMEA

- Approvals

17 SWBS 200 - Propulsion Transmissions & Prime Movers

In the previous section under this SWBS we discussed the propulsors. It is, of course, necessary to twist that devil's tail to actually generate the thrust, so let's talk a little bit about the mechanical side of the problem.

There is little about the marine engineering of AMV propulsion that is unique in principle. The uniqueness comes from the fact that we are often dealing with quite high power levels for the given size of craft, and we tend to have a quite high sensitivity to weight. In some cases we also have a challenge introduced because to the large speed range that must be accommodated, ranging say from 5 to 50 knots.

17.1 Transmitting Power to the Propulsor - AMV Unique Challenges

The transmission system includes the gearboxes, shafts, bearings, etc. The challenges, the tasks this system must face, include concerns with the following:

- RPM - Providing the needed torque at the needed RPM to generate the desired thrust. This is of course common to all marine transmissions, not just those on AMVs.

- Thrust loss (air ingestion) - Waterjet driven craft may have sudden losses of thrust when the waterjet ingests a "gulp" of air. This causes a sudden drop in the torque on the input shaft. This is also very common on race boats, where the boats often leap out of a wave and their propellers race to high RPMs unless controlled.

  The challenge is this drop of torque may cause a spike of RPM to the engine, and then a sudden burst of loading (torque) when the propulsor is re-wetted. These loads are practically impact loads to the transmission system. They can cause damage to gear boxes, and the overspeed potential can destroy engines.

In most cases neither of these problems arise, because there is enough mass in the transmission system to prevent the system from making an instantaneous response to this transient, and the transient is of short enough duration that the load has returned before the system has gotten too far from it's operating point. However, the designer must be aware of the possibility of this type of problem, especially if he is working on a craft with a high likelihood of air ingestion, or with a very low inertia to the drivetrain (e.g, a direct-drive craft which has no gearbox mass to help.)

- Shaft angles - Some AMVs, particularly the surface piercing hydrofoils for example, have a challenge in getting a large propeller far below the hull. This can lead to high shaft angles. These high angles result in non-uniform loading of the propeller, because the blade angle of attack is higher on the downward-moving side of the circle than on the upward moving side. High shaft angles should, of course, be avoided. If they must be used then particular attention must be paid to the propeller design, and the propeller performance will probably be lower than if a lower angle could be adhered to.

  This need not be as strongly the case with surface-piercing props which are designed for high shaft angles, and in some cases this may be enough motivation to select this propulsor.

- Appendage drag - The transmission components that are in the water, e.g. shafts, brackets, etc., do contribute drag. The drag of a spinning shaft can be surprisingly large. This is particularly a problem again in those craft that have long highly-angled wetted shafts, such as some of the surface-piercing hydrofoils. Such craft will also exacerbate the appendage drag issue by having long (and therefore large and thick) shaft struts, usually with enclosed bearings.

- Steerable shafts (e.g. Arneson drives) - Small surface piercing propeller installations, such as the one illustrated earlier, use steerable shafts. There is a universal joint located at the boat's transom, which allows the shaft to swing about 30 degrees port and starboard. This joint is of course subject to wear and of relatively short life. There must also be seals and flexible boots in this area which also become maintenance items.

17.2 RPM Matching & Two-Speed Operations

The biggest design issue in AMV transmission is of course to provide the right RPM to the propulsor. This is nothing AMV-unique, except for ships with a large hump in the drag curve, which may demand a two-speed transmission. Another case is that of an AMV with two distinct operational speeds, say "Cruise" and "Boost". In this case the craft probably sails at Cruise speed by driving one set of propulsors, and then ADDING another set for Boost speed. For example, a trimaran might have two wing jets for Cruise, and two more Main Hull jets for Boost.

This works fine for waterjet craft, since as we saw a waterjet's power absorption does not change with ship speed: Full power = Full RPM, no matter what the speed is. (Of course, full power at low speed may lead to cavitation, but that is a different issue.)

For a propeller however this can be a problem. Recall that Propeller RPM is (approximately) linear with ship speed, a propeller is at its design point when turning X rpm and Y knots, and 2X rpm at 2Y knots. So now imagine that we have a craft with a boost speed of 40 knots and a cruise speed of 20 knots. We design the craft to have propellers for the cruise condition. They are driven by engines that put out the needed cruise power at, say, 2000 rpm. Now when this craft runs at boost speed an additional engine (with its own propeller) is turned on, which adds the extra power needed for boost speed. What happens to the cruise engines? Those propellers will now need to turn at 4000 engine rpm in order to stay at the same J. But 4000 engine rpm is well above the engine limit. What can we do? Let's change the gear ratio so that the cruise engines are only at 2000 rpm at Boost speed. OK, but now they are only at 1000 rpm at cruise speed, and the nature of engines is that at this lower speed they only put out approximately half as much power (a power map for a diesel typical to an AMV is reproduced in Figure 17.1.) Now, while we have bought 1000-horsepower engines, we are only running them at 500-horsepower. This is bad for the engine, and bad for the economics.

What we would like to do is to be able to "shift gears" so that the props can run at 2000 engine-rpm at 20 knots in boost gear, and then we "Shift into second" so that prop rpm doubles for the same engine rpm. In that case we will be at 2000 engine-rpm at 40 knots, at double the prop-rpm.

17.2.1 Two Speed Gearboxes from ZF-Marine

A limited number of 2-speed gearboxes do exist on the market, made by ZF marine. They are only for small engines, and they do not have the 2:1 range of ratios that I described in my example, but they are nevertheless

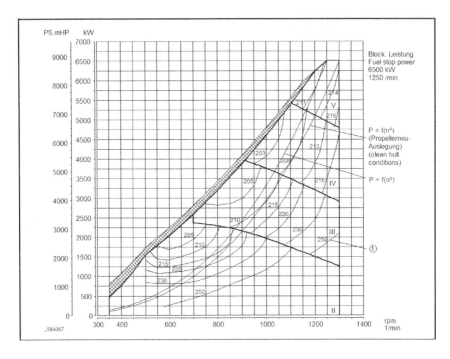

Figure 17.1: A typical AMV diesel engine power map

interesting, and are a development that deserves to be watched in the future.

17.2.2 Waterjets in Two-Speed Applications

The other solution to the two-speed problem is to use waterjets as the two-speed propulsor. Recall that WJ RPM doesn't change with speed, only with power. This means that the waterjet will absorb full power at 20 knots at one RPM, and full power at 40 knots at *almost* the same RPM.

One of the most successful two-speed designs I have personal familiarity with was on motoryacht GENTRY EAGLE. She cruised at a leisurely 47 knots on her diesels, driving KaMeWa waterjets, and when boost speed was needed an Arneson surface-piercing drive was lowered into the water, driven by a gas turbine, yielding a top speed of 61 knots.

360

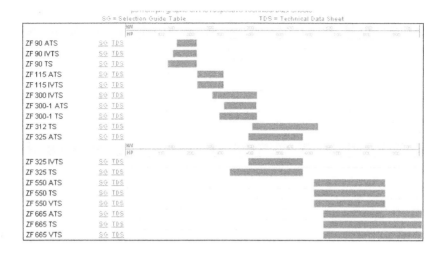

Figure 17.2: Two-speed gearboxes available from ZF Marine

17.3 Prime movers and their selection

We have discussed gear and propulsors. The prime movers form the real source of the power. There is very little that's AMV-unique, except for the weight sensitivity. AMV designers concern themselves assiduously with minimizing the weight of the ship components. This means that an AMV designer is far more likely to be familiar with the high-speed / light weight engines than with, say, Low Speed Diesels.

Of course, the lighter weight engines, including high speed diesels and gas turbines, have higher fuel consumptions than low-speed engines. So at some point, when range becomes large, the weight of fuel becomes the dominant factor in the tradeoff and the architect picks a heavier engine in order to reduce total ship displacement.

Homework assignment: For a given ship characteristic, estimate the weight of the engine and the fuel, assuming (a) an MTU diesel, and (b) a gas turbine, for two different range cases. Equations for fuel consumption to be given. Below some range, the sum "Engine -plus- Fuel" is lighter for the gas turbine, and then above that range the sum favors the diesel. For a homework-given set of engine parameters, what is this cross-over range?

Light Duty
2 Speed   L

MODEL	RATIOS		POWER/RPM		MAXIMUM RATED POWER						MAX RPM	WEIGHT		BELL HSOS. AND NOTES
	1st	2nd	kW	hp	2100 rpm		2300 rpm		2450 rpm			kg	lb	
					kW	hp	kW	hp	kW	hp				
ZF 665 ATS	1.525	1.225,1.277,1.325,1.363,1.401	0.3729	0.5000	783	1050	858	1150	914	1225	2500	353	777	SAE 1
10 degrees	1.757	1.411,1.471,1.526,1.571,1.613												
	1.971	1.583,1.650,1.712,1.762,1.810												
	2.226	1.788,1.863,1.934,1.990,2.044	0.3654	0.4901	767	1029	841	1127	895	1201	2500			
	2.448	1.966,2.050,2.127,2189,2.248												
	2.517	2.022,2.107,2.187,2251,2.312	0.3566	0.4783	749	1004	820	1100	874	1172	2500			
	2.960	2.377,2.478,2.572,2646,2.718	0.2917	0.3912	613	822	671	900	715	958	2500			
ZF 665 TS	1.111	0.892,0.930,0.965,0993,1.020	0.3729	0.5000	783	1050	858	1150	914	1225	2500	344	757	SAE 1
	1.182	0.949,0.989,1.027,1.057,1.085												
	1.262	1.013,1.056,1.096,1128,1.159												
	1.400	1.124,1.172,1.216,1252,1.286												
	1.500	1.205,1.256,1.303,1341,1.378												
	1.743	1.400,1.459,1.514,1558,1.601												
	2.000	1.606,1.674,1.738,1788,1.837												
	2.233	1.794,1.870,1.941,1997,2.051	0.3654	0.4901	767	1029	841	1127	895	1201	2500			
	2.593	2.082,2.171,2.253,2318,2.381												
	3.042	2.443,2.547,2.643,2719,2.793	0.2917	0.3912	613	822	671	900	715	958	2500			
ZF 665 VTS	1.525	1.225,1.277,1.325,1.363,1.401	0.3729	0.5000	783	1050	858	1150	914	1225	2500	158	348	SAE 1
10 degrees	1.757	1.411,1.471,1.526,1571,1.613												
	1.971	1.583,1.650,1.712,1762,1.810												
	2.226	1.788,1.863,1.934,1990,2.044	0.3654	0.4901	767	1029	841	1127	895	1201	2500			
	2.448	1.966,2.050,2.127,2189,2.248												
	2.517	2.022,2.107,2.187,2251,2.312	0.3566	0.4783	749	1004	820	1100	874	1172	2500			
	2.960	2.377,2.478,2.572,2646,2.718	0.2917	0.3912	613	822	671	900	715	958	2500			

* Special Order Ratio.
11:50 AM GMT - 04-Apr-08

Figure 17.3: Gear ratios available on the ZF two-speed gears

18 SWBS 200 - Breguet's Range Equation

The conventional range calculation for a surface ship is very simple, and (stripped of various complicating factors) it may be expressed as:

Fuel weight = Fuel Burn Rate x time en route

Fuel burn rate = SFC x SHP

Time en route = Range / Speed

The conventional range calculation thus depends only upon SFC, the ship's resistance (SHP), the range, and the speed. We then in practice treat all of those parameters as constant: Speed is constant across the miles traveled, power is constant, SFC is constant.

These three assumptions are basically true only if displacement is also constant. Certainly if the displacement changes, then the resistance should change, no? And if the ship burns off her fuel as she travels, then must not the displacement change? There are only two times when this would not be true: If the fuel burn is so small as to be a negligible change in weight, or if the ship takes on ballast continuously during her transit in order to maintain weight the same.

In the case of long-ranged Advanced Marine Vehicles we do not make this constant-displacement assumption, and we invoke a different means of calculating range: The Breguet range equation.

Let us work our way toward the Breguet equation by first re-writing the conventional range calculation in terms of Lift/Drag ratio (inverse Drag/Weight ratio), Propulsive Coefficient, and Weight. The result is:

$Range = 198e3 * OPC * \frac{W_{fuel}}{W_{total}} * \frac{L}{D} * \frac{1}{SFC}$

Where:

Range in nautical miles

$198e3$ = (grams / tonne) (knots / meters-per-second) / (g=9.8 m/s2)

W_{fuel} = Weight of fuel (tonnes)

W_{total} = Weight of ship (tonnes)

OPC = Overall Propulsive Coefficient

L/D = Ship Lift-to-Drag ratio

SFC = Specific Fuel Consumption (grams / kW-hour)

The Breguet Range equation was developed in aviation engineering - since airplanes don't take on ballast. The derivation is easily found in an internet serach. The following presents a marinized version of Breguet's formula, compared to the constant-displacement formula given above.

Constant Displacement formula:

$$Range = 198e3 * OPC * \frac{W_{fuel}}{W_{total}} * \frac{L}{D} * \frac{1}{SFC} \tag{18.1}$$

Variable Displacement (Breguet) formula:

$$Range = -ln(1 - \frac{W_{fuel}}{W_{total}}) * 198e3 * OPC * \frac{L}{D} * \frac{1}{SFC} \tag{18.2}$$

The Breguet formula, with the introduction of one logarithmic term, captures the fact that the ship gets lighter as fuel is burned off. Instead of assuming constant displacement, this formula instead assumes constant L/D.

Of course, this is simply the replacement of one assumption by another, and is still subject to verification on any given project. While it's true that the large GM of some multihulls means that ballast is not required, in the case of a trimaran perhaps the change in displacement will cause the Amas to come out of the water, requiring ballast to keep them immersed? Also, on some hulls the Drag/Weight ratio may not in fact be constant.

If these assumptions are valid, then the effect of the Breguet calculation can be dramatic. Table 18.1 presents a calculation of the impact of this effect, for varying fuel weight fractions. As may be seen, for very large fuel fractions ($> 80\%$ Full Load - which is unlikely!) the Breguet effect amounts to a doubling of the range. At smaller fractions, say 30-40% Full Load, this still yields a 25% increase in range over the more conventional displacement ship calculation method.

Finally, note that this effect is only realistic if the owner uses it: if he doesn't refuel, and doesn't ballast. The military practice, for example, of never allowing the ship to get below $\frac{1}{2}$ or $\frac{3}{4}$ "tank" will obviate the benefits of this calculation: In effect the owner is running his ship in a Constant Displacement mode, and it behooves the Naval Architect to perform the calculations accordingly.

In conclusion, the Breguet range formula may be an AMV-unique range result, since it requires a hull form that doesn't need ballast (e.g. a catamaran.) The use of the Breguet method may potentially greatly increase the utility of the ship, making possible trans-oceanic passages that would be impossible if the fuel loads were calculated in the conventional manner. Note that this effect is equivalent to a great improvement in SFC or Resis-

$\frac{W_{fuel}}{W_{total}}$	Range (example)	$\ln(1 - \frac{W_{fuel}}{W_{total}})$	Breguet range
0	0	0	0
0.1	100 miles	0.105	105 miles
0.2	200 miles	0.223	223 miles
0.3	300 miles	0.357	357 miles
0.4	400 miles	0.511	511 miles
0.5	500 miles	0.693	693 miles
0.6	600 miles	0.916	916 miles
0.7	700 miles	1.204	1204 miles
0.8	800 miles	1.609	1609 miles
0.9	900 miles	2.303	2300 miles

Table 18.1: The effect of the Breguet range calculation

tance - saving 25% in fuel would require either a breakthrough in engines or hull form, or a simple employment of the Breguet method for range.

But that employment rests in the hands of the ship owner. The assumptions embedded in the Breguet formula are subject to violation by an uninformed operator.

19 SWBS 500 - Lift Fan Systems

SWBS 500 is the accounting group for ship auxiliary systems. This includes the "normal" auxiliary systems such as firefighting, sewage, air conditioning, etc. about which I have nothing to say. But SWBS 500 is also where we include the lift fan system for the air cushion vehicles and SES, which will form the subject of this unit.

There are two basic steps to the lift system design: We must estimate the amount of flow and pressure that are required for the ship, and then we must design a fan suite that delivers that flow, at that pressure.

Our lectures on this subject will provide some understanding of the air demands of this type of cushion, and the characteristics of fans that will supply this air.

19.1 Cushion Air Demand - Estimating P & Q

"P" and "Q" are the conventional symbols for the air flow (Q) and pressure (P) in a powered-sustention air cushion AMV. The pressure is, of course, determined by the hydrostatic balance as discussed earlier. The weight of the craft is borne by the cushion pressure acting on the cushion area, plus any contribution from sidehull buoyancy. Thus finding "P" is easy, the challenge is to find the design value of the flow, "Q." I shall cover three methods for estimating air flow demand: Similitude from previous ships, the "hovergap" method, and the "wave pumping" method. In practice all three are used in various combinations, as will also be shown.

19.1.1 Air Flow Similitude

In practice, designers will collect data on successful vessels and will use this to form guidance. To this end, the first thing that we need is a scaling relationship that will allow us to take the flow from one ship and use it to estimate the flow on another ship. The estimating relationship is as follows:

$Q = \bar{Q} * S_c * \sqrt{(2P_c/\rho_{air})}$

Where:

S_c =cushion area $(L_c * B_c)$

P_c =cushion pressure

ρ_{air} =air density

Note that the term $\sqrt{(2P_c/\rho_{air})}$ yields an air exit velocity (e.g. m/sec)

$\bar{Q}$ is a non-dimensional flow parameter that is extracted from parent craft or from experiment. Yun & Bliault offer a wide guidance band, as follows:

ACV: $\bar{Q} = 0.015 - 0.050$

SES: $\bar{Q} = 0.005 - 0.010$

19.1.2 The Hovergap Method for Air Demand

The hovergap method is a static model of the air flow situation - it yields a time-invariant value for Q. This method states that the craft may be considered to hover above the sea surface with some measurable hovergap, through which air will flow. The velocity of flow through an orifice is:

- $V = \sqrt{(2P_c/\rho_a)}$

So if we know the height and the permieter, then we can calculate the area of this orifice. Area times velocity equals flow, thus:

- h = hovergap (height)

- L = Perimeter (length)

- Q = h x L x V

- Q = h x L x $\sqrt{(2P_c/\rho_a)}$

The challenge therefore is obviously to have an estimate of the hovergap. The practical solution is to scale it from a parent craft, as:

- $Q_2 = Q_1 \frac{L_2}{L_1} \sqrt{\frac{P_2}{P_1}}$

One practitioner provided me with the table of data given in Figure 19.2. This provides useful data on a number of ACVs (hovercraft). What is interesting in this data set is to plot the flow parameter, as a function of P and L. This has been done in Figure 19.3, in which case the "Flow Parameter" is simply $(Perimeter x \sqrt{(Pressure)})$. As may be seen, the data suggests that there is strong dependency upon the service speed of the craft, with hoverbarges and other low-speed ACVs having one trend line, and the fast ACVs having a very different one. The inverse slope of the trend line yields the CFS of flow per unit $(Perimeter x \sqrt{(Pressure)})$.

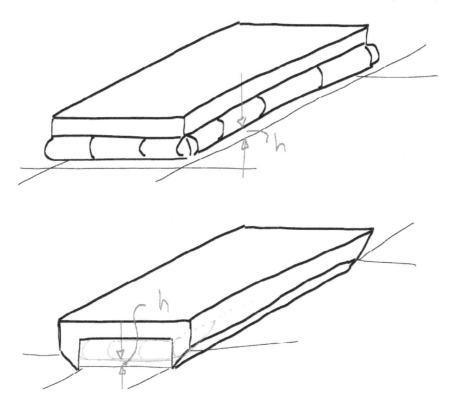

Figure 19.1: Stylized illustration of the hovergap for an ACV (top) and an SES (bottom)

19.1.3 Wave Pumping

Up to this point we have treated the air demand as a quasi-static problem, dependent only on pressure and size. But the data in Figure 19.2 hints to us that there is a dynamic dependence to this too. In this next unit we will consider a totally dynamic approach to modeling air flow demand. The method is called "wave pumping" and is an attempt to model the cushion air demand as if the cushion were a volume that is continuously being "pumped" by the ocean waves. Figure 255 is a crude sketch drawn by me that shows an SES cushion profile, with the bow and stern skirts visible at the ends. In red are shown two positions for a passing wave, one when the crest is amidships, and one when the trough is amidships. These two conditions give rise to a change in cushion volume that took place between time of the passage of the crest to the trough, and this volume must be

Craft	Length-FT	Beam-FT	Weight- LB	Flow-CFS		Cushion Pressure-PSF
TAV-40	75	28	146000	800		70
PUC-22	110	42	163000	800		35
MDRIC	63	32	180200	1000		89
PACK	80	32	320000	1960		125
ACT 100	80	61	500000	1710		102
A-200	130	62	783000	2790		97
YP I	127	84	828000	2217		78
TM160	129	84	865000	3066		80
YP II	126	81	923000	2200		90
Sea Pearl	180	80	1680000	4500		117
LACV-30	76	36	116000	3243		42
Jeff B	80	40	324000	10000		101
LCAC	80	40	340000	7800		106
BAC 150	80	53	349440	12200		82
N 500	130	70	593600	17000		65
SRN 4	185	78	627000	16000		43

Figure 19.2: Air flow demand data for a collection of hovercraft

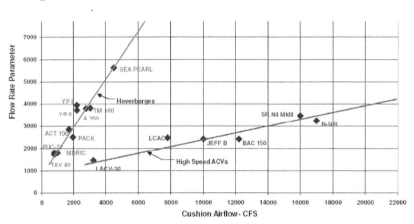

Figure 19.3: The data from Figure 19.2, plotted showing an apparent sensitivity of Flow to Speed

refilled with air by the lift fans. In effect, the waves act as pistons in an air pump, hence the term "wave pumping."

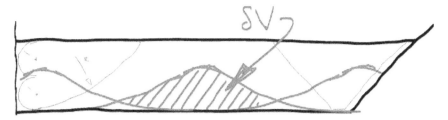

Figure 19.4: A crude sketch of an SES profile, showing the volume of the cushion that must be refilled with air between the passage of a crest and a trough.

The velocity with which this volume changes, the rate of change of the volume, may be thought of as: $\frac{dVol}{dt} = f$(wave height, wave length, encounter speed)

For realistic conditions this can be written as:
$\frac{dVol}{dt} = -B_c H v sin(a_r)$
Where:

- B_c = cushion beam

- H = wave height

- $a_r = -\pi L_c/\lambda$

- v = speed relative to the wave = $V_s \pm gT_0/2\pi$

- L_c = cushion length

- λ = wave length

- T_0 = wave period

This in turn is equal to: $dVol/dt = -B_c H(V_s \pm gT_0/2\pi)sin(-\pi L_c/\lambda)$ Now, we don't actually care about the dVol/dt - we are trying to decide how big the fans have to be on the boat. So what we care about is the maximum value. The sine term will obviously maximize at 1.0, so that maximum value of dVol/dt becomes: $dV/dt(max) = -B_c H(V_s \pm gT_0/2\pi)$

In practice we don't need to size the fan to the instantaneous maximum, we size it for a flow of about 35% of that maximum. This gives rise to the wave pumping design formula:
$Q_{design} = 0.35 B_c H(V_s + gT_0/2\pi)$

19.2 Air Demand > Air Supply

Now that we have estimated the air flow required, let us see what sort of fan will provide that needed air. We have seen that, due to wave pumping, the air demand has time-varying characteristics, in which it varies more or less at wave period. During these variations the craft weight doesn't change, so we still want the same cushion pressure at all points in the wave-pumping cycle. This means that the ideal lift fan would be one that delivers a constant pressure across some range of flow. It would have a P / Q characteristic that is flat, as sketched in Figure 19.5. Unfortunately, real fans have P/Q characteristics that are humped, as in Figure 19.6.

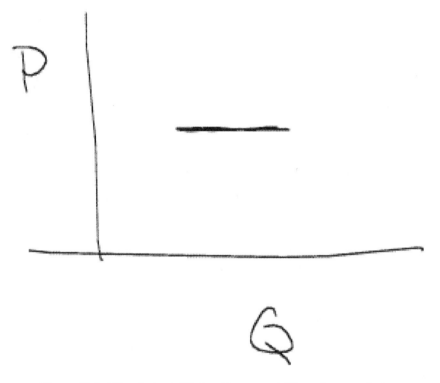

Figure 19.5: The desired lift fan Pressure / Flow characteristic

Figure 19.7 presents the P / Q characteristic for the Howden Buffalo L-25 fan, used in one SES design project of my experience. The design point for the SES, marked in pencil on the fax, corresponds to a flow of about 30,000 cfm (cubic feet per minute) per fan, at a pressure of about 55 inches

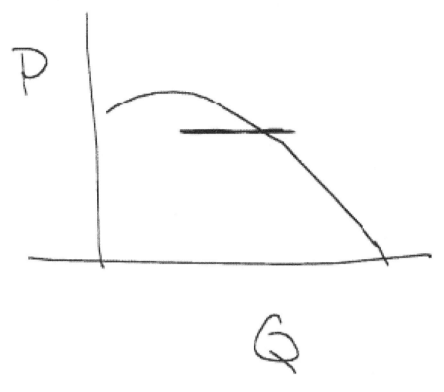

Figure 19.6: The shape of a real fan's pressure / flow characteristic

of water. There is a flat part on the L25 fan curve, located at about 15,000 cfm. Why don't we operate the fan there, where the pressure won't change much with variations in flow? The answer is because of what happens to the left of this region on the curve. To the left is where the fan goes into stall. Consider a "walk" along the fan curve from right to left. Imagine that this fan is powering a shop vac or similar blower, and we are going to throttle the flow by putting our hand over the hose. As we lower the flow (as we move right to left on the curve) the pressure goes up - and we feel increasing resistance on our hand. But as we pass the peak of the curve all of a sudden the pressure goes down as we choke the flow. In a real shop-vac this drop in pressure at the last inch is quite noticeable, and one can hear the motor rpm change as the fan wheel goes into stall and the power drops way off. If this happens in an SES it means that the cushion pressure drops off, and this means that the craft is all of a sudden not an SES, but will drop into catamaran mode. It is much more beneficial to have some degree of slope to the fan curve, such that when a wave crest arises (and flow drops off) this yields a rise in cushion pressure which will help lift the craft higher in the water and thus across the wave.

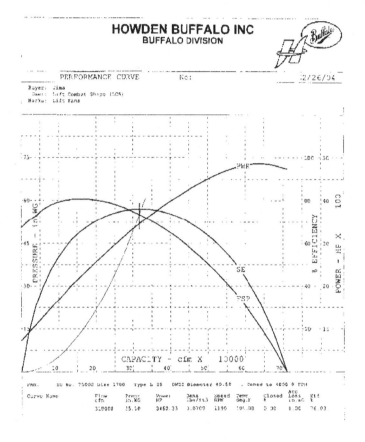

Figure 19.7: A real SES lift fan. The curve for "FSP" is the fan static pressure in inches water gage, plotted versus the flow in cfm x 10,000. Other curves give efficiency and power consumed by this fan.

19.3 Fans 101

At this point we seem to have moved from talking about lift demand to talking about fans, so let's study fan aerodynamics a little bit. In classroom lectures on this unit I illustrate this with slides I obtained from a course in refrigerant-cycle air conditioning from Syracuse University. I wish to highlight this fact because it underscores that as AMV designers we will find ourselves drawing from fields that are not traditionally naval architecture - such as fan design. In an air conditioning plant there are two primary places that fluid-movers are found: On the refrigerant side, in the form of compressors and pumps, and on the air-handling side, in the form of fans. Fluid-movers on both sides of the problem may be classed into two categories. Again, taking notes from Syracuse University (see Figure 19.8) they may be classed as:

- Positive displacement machines - such as hydraulic pumps and motors

- Roto-Dynamic machines - such as gas compressors, turbines, windmills, propellers, and fans

The roto-dynamic machines in turn are classed as either Axial flow, Centrifugal flow, or Mixed flow. Note that "radial flow" is a synonym for "centrifugal flow." The Syracuse slides include Figure 19.9 which attempts to depict the difference between axial and centrifugal flow machines. Compare this to the similar illustration under "waterjets" in Chapter 16.

When discussing axial devices, I particularly appreciate the irony of a mechanical engineer showing his class pictures of ship propellers, (Figure 19.10), while I show air conditioning machinery to a room full of ship designers. Physics is physics - it is only that we have chosen to employ that physics to serve different aims. Figure 19.11 continues the series.

What this amounts to is that a fan designer has a range of types of machine from which to choose. To make his choice he characterizes the desired performance of the fan. Fan performance is characterized by the following parameters:

- Pressure rise (head) expressed in units such as inches of water (1 inch w.g. = 5.204 lb/sq.ft)

- Volumetric flow rate expressed in units such as cfm

- Rotational speed expressed in units such as rpm

- Fan fluid power (the energy imparted to the fluid) expressed in units such as horsepower

- Fan shaft power expressed in units such as horsepower

- Fan efficiency (fluid power divided by shaft power) - dimensionless

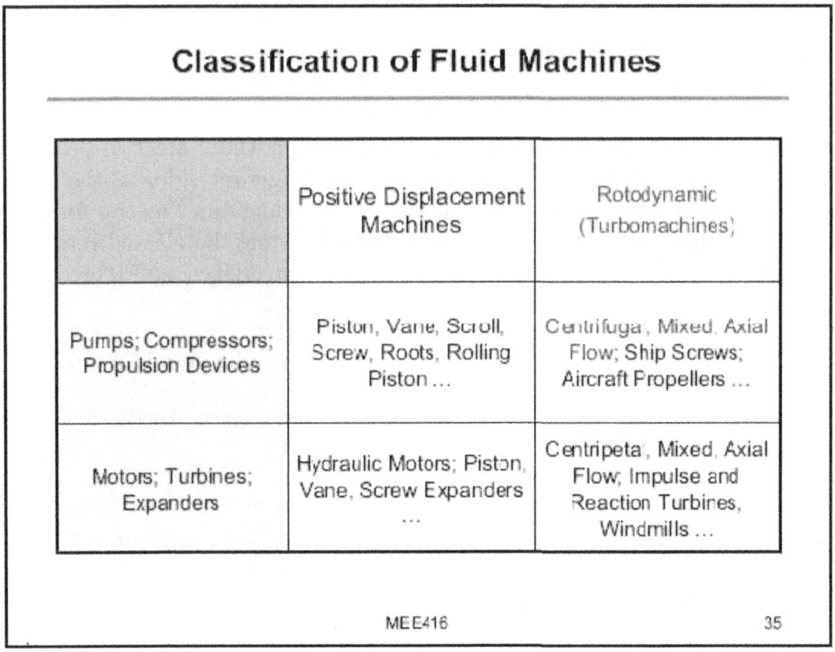

Classification of Fluid Machines

	Positive Displacement Machines	Rotodynamic (Turbomachines)
Pumps; Compressors; Propulsion Devices	Piston, Vane, Scroll, Screw, Roots, Rolling Piston ...	Centrifuga, Mixed, Axial Flow; Ship Screws; Aircraft Propellers ...
Motors; Turbines; Expanders	Hydraulic Motors; Piston, Vane, Screw Expanders ...	Centripeta, Mixed, Axial Flow; Impulse and Reaction Turbines, Windmills ...

MEE416   35

Figure 19.8: Syracuse University slide on the types of Fluid Movers

Fan performance is presented as either tables or charts, showing the pressure rise (Δ P), efficiency (η), and power (W) as a function of the volume flow rate (Q) for different speeds (RPM.) We have seen such curves in the case of the Howden Buffalo L25, previously. The fluid power (Wf) is very important to understand. It is also quite simple to calculate. The fluid power is the useful power imparted to the fluid by the fan. It is given by $Wf = \Delta P * Q$

The shaft power is the total mechanical power delivered to the fan by the shaft, and it is greater than the fan power. Efficiencies for well designed lift fans are generally somewhere in the neighborhood of 50%.

Vendors have a wide range of choices for fans. Figure 19.12 illustrates the Howden Buffalo commercial fan range. Note that the scales on this graph are logarithmic: They make fans that cover four orders of magnitude in flow, and five orders of magnitude in pressure.

Once the naval architect has converged the ship air flow demand, he can rest fairly confident that a commercial fan can be found to provide this flow.

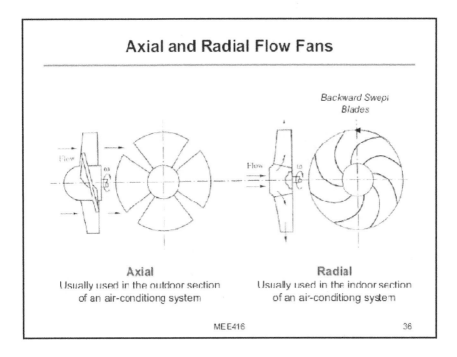

Figure 19.9: Depiction of the difference between axial and centrifugal aeromachinery

19.4 Fan Scaling Laws

During the early parametric stages of an SES design the naval architect frequently needs to perform her own fan sizing estimates, usually by scaling from other existing fans. There are two most important dimensionless coefficients which describe a fan's performance. These are the Flow Coefficient and the Pressure Coefficient. They are defined as follows:

Pressure coefficient: $\Psi = \Delta P/(\rho N^2 D^2)$

Flow coefficient: $\Phi = Q/ND^3$

Where:

N = fan RPM

D = fan diameter

From these parameters we can derive fan scaling laws.

A fan's P/Q/RPM map can be redrawn in terms of Ψ and Φ. When it is redrawn in that manner it becomes "generic" in the sense of being able to be scaled to any desired size. Figure 19.13 shows an illustration of the P/Q

Figure 19.10: A mechanical engineer's illustration of two axial flow machines

curves for a given fan design, at two different sizes and rpms. When they are re-plotted in terms of Φ and Ψ the two fans "collapse" and are revealed to be the same fan - Figure 19.14. (The errors in the illustrated case are because this is experimental data measured in the classroom in Syracuse, as a homework problem to make exactly this point.)

In many practical cases we don't actually re-plot the fan curve in non-dimensional terms and then re-scale to a new size. If we know the scaling that we want, then we can employ these non-dimensional relationships to yield a set of scaling laws as follows. In each case the subscripts 1 and 2 refer to taking Fan-1 and scaling it to a new size to yield Fan-2. Fan Laws

- $Q_2 = Q_1 x(N_2/N_1)x(D_2/D_1)^3$

- $H_2 = H_1 x(N_2/N_1)^2 x(D_2/D_1)^2 x(\rho_2/\rho_1)$

- $HP_2 = HP_1 x(N_2/N_1)^3 x(D_2/D_1)^5 x(\rho_2/\rho_1)$

Fan scaling equations

Figure 19.11: This turbocharger shaft shows two mixed-flow machines, one (the turbine) to extract energy from the exhaust gas and the other (the compressor) to impart energy into the inlet flow

- $(D_2/D_1) = ((Q_2/Q_1)^2/(P_2/P_1))^{0.25}$
- $(N_2/N_1) = (Q_2/Q_1)/(D_2/D_1)^3$

Horsepower

- HP = 1.340 x cms x kPa / efficiency
- kW = cms x kPa / efficiency

An example of this type of scaling is shown in Table 16, wherein we took three "parent" fans, designated "Chinese" "HLCAC" and "Skjold" and we attempted to scale them to our design case of 18.63 kPa and 200 cms. As may be seen the three different parents yielded three different offspring fans. Note finally the fact that in this case we calculated the tip speed of the fan, being defined as the revolutions per second, times the diameter, times pi. The tip speed should be maintained below the speed of sound by a good margin, in this case using a limit of 600 feet per second.

	Chinese	HLCAC	Skjold
Parent P	6.03 kPa	7.94 kPa	8.34 kPa
Parent Q	200 cms	148.7 cms	75 cms
Parent diameter	3.0 meters	1.6 meters	1.3 meters
Parent speed	700 rpm	1692 rpm	1800 rpm
Efficiency	84%	68.4%	80%
Scaled diameter	2.26 meters	1.5 meter	1.736 meters
Scaled rpm	1631 rpm	2764 rpm	2015 rpm
Physical Size	Unknown	77.1"W, 116.6"H,98"L	Unknown
Scaled tip speed	634 ft/s. Higher than recommended 360 ft/s.	711 ft/s	601 ft/s
Comments	H-bar = 0.14. efficiency < 65% . Very high risk	Tip speed too high (should be less than 600 ft/s)	Possibility. Tip speed high for design. Scaling risk

Table 19.1: Three different parent fans all scaled to $P = 6.03$ kPa & $Q = 200$ cms

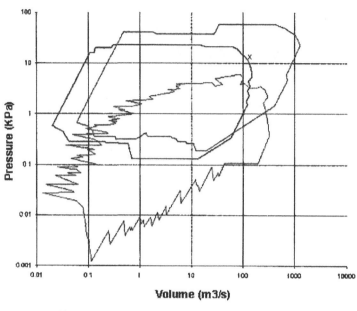

Figure 19.12: Howden Buffalo fan product ranges

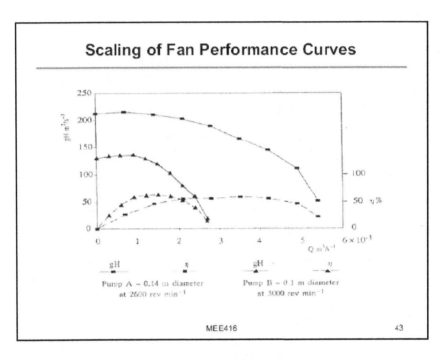

Figure 19.13: A given fan design, in two different sizes to yield two different P/Q curves

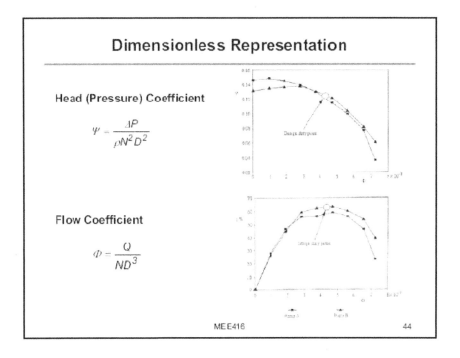

Figure 19.14: The same two fans as in Figure 19.13, but when plotted non-dimensionally revealed to be the same turbomachine

20 About the Author

The author is a naval architect with 35 years of experience, primarily in unconventional vehicles and projects. He holds a 2013 PhD in Engineering and Applied Science from the University of New Orleans, in turn founded upon a 1979 BSE in Naval Architecture and Marine Engineering from the University of Michigan, and a 2010 Master's degree from UNO. He is licensed as a Professional Engineer in Naval Architecture and Marine Engineering in the State of Washington.

The author's career history may be reviewed by visiting his consultancy website at *mckesson.us*. As is therein shown, he has spent most of his career as a design and consulting engineer, but has recently turned to teaching in order to pass to a new generation his passion for ship design. McKesson has a long personal interest in Advanced Marine Vehicles, dating back at least to his very first job offer after college, which was in the Navy's 3KSES program. (Actually, McKesson's interest goes back even further, to the early 1970s and boyhood days crawling around a Russian hydrofoil pleasure boat which was then being (unsuccessfully) imported by Kettenburg Boat Builders in San Diego California.)

Declining the 3KSES job offer McKesson has nevertheless had many interesting positions and projects over the years, and many wonderful opportunities to work with real luminaries in this field. Through that career he has learned from these leaders, and is pleased to here offer a summary of their wisdom for the benefit of a further generation.

Bibliography

[1] G. Gabrielli and T. von Karman, "What price speed - specific power required for propulsion of vehicles," *Mechanical Engineering*, vol. 72, October 1950. pp 775-778.

[2] C. G. Kennell, "Design trends in high speed transport," *Marine Technology*, vol. 35, no. 3, 1998.

[3] M. Templeman and C. Kennell, "The effect of ship size on transport factor properties," in *Proc. First International Conference on High-Performance Marine Vehicles (HIPER'99)*, pp. 210–219, 1999.

[4] O. Faltinsen, *Hydrodynamics of High-Speed Marine Vehicles*. Cambridge University Press, 2006.

[5] E. O. Tuck and L. Lazauskas, "Unconstrained ships of minimum total drag," *self published*, 2008.

[6] H. Lindgren and Å. Williams, *Systematic Tests with Small, Fast Displacement Vessels: Including a Study of the Influence of Spray Strips*. Meddelanden från Statens Skeppsprovningsanstalt, Akademiförlaget-Gumperts, 1969.

[7] E. A. Butler, "The Surface Effect Ship," *Naval Engineers Journal*, vol. 97, no. 2, pp. 200–253, 1985.

[8] L. Yun and A. Bliault, *Theory & Design of Air Cushion Craft*. Elsevier Science, 2000.

[9] J. Allison, *Marine Waterjet Propulsion*. Society of Naval Architects and Marine Engineers, Transactions, 1993.

[10] NAVSEA, *ESWBS: Expanded Ship Work Breakdown Structure*. US Navy, 1985.

[11] D. J. Clark, W. M. Ellsworth, and J. R. Meyer., "The quest for speed at sea," *Technical Digest*, April 2004.

[12] C. G. Kennell, *Technical and Research Bulletin 7-5: SWATH Ships*. The Society of Naval Architects and Marine Engineers, January 1992.

[13] K. V. Rozhdestvensy, "Wing-in-ground effect vehicles," *Progress in Aerospace Sciences*, vol. 42, November 2006.

[14] W. J. Egginton and N. Kobitz, *The Domain of the Surface-Effect Ship*, pp. 268–298. Society of Naval Architects and Marine Engineers, Transactions, 1975.

[15] H. Söding, "Drastic resistance reductions in catamarans by staggered hulls," *The 4th International Conference on Fast Sea Transportation, FAST 1997*, 1997.

[16] P. Kamen, B. Duffy, and C. Barry, "Ferries for the San Francisco Bay area; new paradigms from new technologies," *Published on line at: http://www.well.com/user/pk/waterfront/Ferry/ferry-pk-020604.html*, 2004.

[17] V. D. Norman, "Speed and transport economy," in *Conference on High Speed Craft*, (Kristiansand Norway), Norwegian Society of Chartered Engineers, 1994.

[18] K. S. Davidson, "Notes on the power-speed-weight relationship for vehicles," *Journal unknown*, March 1951.

[19] O. K. Ritter and M. T. Templeman, *High-speed sealift technology*, vol. CDNSWC-TSSD-98-009. Carderock Division - Naval Surface Warfare Center, 1998.

[20] C. G. Kennell, "On the nature of the transport factor component TF-ship," *Marine Technology*, vol. 38, no. 2, 2001.

[21] C. B. McKesson, "A parametric method for characterizing the design space of high speed cargo ships," *The Royal Institution of Naval Architects Conference on High Speed Craft, ACVs, WIGs and Hydrofoils*, 2006.

[22] H. Benford, *The practical application of economics to merchant ship design*. No. 012 in University of Michigan Department of Naval Architecture and Marine Engineering, University of Michigan Department of Naval Architecture and Marine Engineerings, 1976.

[23] M. A. Redmond, "Ship weight estimates using computerized ratiocination," in *43d Annual Conference*, (Atlanta, GA), Society of Allied Weight Engineers, May 1984.

[24] C. B. McKesson, "A collection of simplified field equations for surface effect ship design," in *Intersociety Advanced Marine Vehicles Symposium*, 1992.

[25] L. J. Doctors, V. Tregde, C. Jiang, and C. B. McKesson, "Optimization of a split-cushion surface-effect ship," in *FAST 2005 - Eight International Conference on Fast Sea Transportation*, 2005.

[26] E. O. Tuck and L. Lazauskas, "Optimum hull spacing of a family of multihulls," *self published*, 1998.

[27] M. Gertler, *A Reanalysis of the Original Test Data for the Taylor Standard Series*. Navy Department, 1954.

[28] H. Y. H. Yeh, "Series 64 resistance experiments on high speed displacement forms," *Marine Technology*, July 1965.

[29] J. M. Zips, "Numerical resistance prediction based on the results of the VWS hard chine catamaran hull series," *The 3rd International Conference on Fast Sea Transportation, FAST 1995*, 1995.

[30] A. Molland, S. Turnock, and D. Hudson, *Ship Resistance and Propulsion: Practical Estimation of Propulsive Power*. Cambridge University Press, 2011.

[31] A. S. Toby, "To the edge of the possible: US high speed destroyers, 1919–1942 part 2: Secondary hull form parameters," *Naval Engineers Journal*, vol. 114, no. 4, pp. 55–76, 2002.

[32] M. Insel, *An investigation into the resistance components of high speed displacement catamarans*. Original typescript, 1990.

[33] T. Armstrong, "The effect of demihull separation on the frictional resistance of catamarans," *The 7th International Conference on Fast Sea Transportation, FAST 2003*, 2003.

[34] I. Mizine, G. Karafiath, P. Queutey, and M. Visonneau, "Interference phenomenon in design of trimaran ship," in *FAST 2009 - Tenth International Conference on Fast Sea Transportation*, 2009.

[35] L. J. Doctors and C. B. McKesson, "The resistance components of a surface effect ship," in *Proceedings of The Twenty-Sixth Symposium on Naval Hydrodynamics*, 2006.

[36] J. N. Newman and F. A. P. Poole, "The wave resistance of a moving pressure distribution in a canal," *Schiffstechnik*, 1962.

[37] L. J. Doctors and S. D. Sharma, "The wave resistance of an air cushion vehicle in steady and accelerated motion," in *Journal of Ship Research*, vol. 16:4, 1972.

[38] J. Schuler, ed., *Modern Ships and Craft: Ed.* Naval Engineers Journal, American Society of Naval Engineers, 1985.

[39] D. Kelsall, "Resistance of catamarans," *Professional Boatbuilder*, April/May 2011.

[40] H. Saunders and R. Taggert, *Hydrodynamics in ship design.* Hydrodynamics in Ship Design, Society of Naval Architects and Marine Engineers, 1965.

[41] P. Mantle, *Air Cushion Craft Development. First Revision.* Defense Technical Information Center, 1980.

[42] G. R. Lamb, "Some guidance for hull form selection for swath ships," *Marine Technology*, October 1988.

[43] I. W. Dand, "High speed craft bow diving in following seas," *The International Conference High Speed Craft – ACV's, WIG's & Hydrofoils*, 2006.

[44] S. Steen, *Cobblestone Effect on SES.* Norwegian Institute of Technology, 1993.

[45] NATO, *Standardization Agreement (STANAG): Subject : Common Procedures for Seakeeping in the Ship Design Process.* North Atlantic Treaty Organization, Military Agency for Standardization, 2000.

[46] J. F. O'Hanlon and M. E. McCauley, "Motion sickness incidence as a function of the frequency and acceleration of vertical sinusoidal motion," *Aerospace Med*, 1974.

[47] RINA, *Transactions of the Royal Institution of Naval Architects.* No. v. 137, Royal Institution of Naval Architects, 1995.

[48] A. Blyth, "SES stability in turns – the influence of sidewall shape," *International High Performance Vehicle Conference*, 1988.

[49] NAVSEA, *DDS 079-1: Stability and Buoyancy of U.S. Naval Surface Ships.* US Navy.

[50] S. Davis and A. Malakhoff, "Transversely stiffened membrane seal," June 8 1982. US Patent 4,333,413.

[51] J. C. Rose and C. Kruppa, "Surface piercing propellers – methodical series model test results," *The First International Conference on Fast Sea Transportation, FAST 1991*, 1991.

[52] J. C. Rose, C. Krupp, and K. Koushan, "Surface piercing propellers – propeller hull interaction," *The Second International Conference on Fast Sea Transportation, FAST 1993*, 1993.

Made in United States
Orlando, FL
10 January 2025